SECRETS OF THE

And like as rigour of tempestuous gusts
Provokes the mightiest hulk against the tide
To suffer shipwreck
    Shakespere, Henry IV, Act 5

There is a tide in the affairs of men,
Which, taken at the flood, leads on to fortune;
Omitted, all the voyage of their life
Is bound in shallows and in miseries.
On such a full sea we are now afloat;
And we must take the current when it serves,
Or lose our ventures.
    Julius Caesar, Act 4

SECRETS OF THE TIDE
Tide and Tidal Current Analysis and Applications, Storm Surges and Sea Level Trends

JOHN D. BOON
Professor Emeritus
Department of Physical Sciences
Virginia Institute of Marine Science
Virginia, USA

Horwood Publishing
Chichester, U.K.

HORWOOD PUBLISHING LIMITED
International Publishers in Science and Technology
Coll House, Westergate, Chichester,
West Sussex PO20 3QL England

Published in 2004

© J.D. Boon. All Rights Reserved. No part of this publication may be reproduced, stored in a retrieval system, or transmitted in any form or by any means, electronic, mechanical, photocopying, recording, or otherwise, without the permission of Horwood Publishing

ISBN: 1-904 275-17-6
British Library Cataloguing in Publication Data
A catalogue record of this book is available from the British Library

Printed by Antony Rowe Limited, Eastbourne

Table of contents

Preface viii

1. Tidewater 1

    1.1 Tidewater Geography 1
    1.2 Early Use of Tidal Power 2
    1.3 The Intertidal Zone 3
    1.4 Tidal Wetlands 4
    1.5 Tidal Circulation and Tidal Mixing in Estuaries 5
    1.6 Effects of Stratification – Hypoxia and Anoxia 7
    1.7 Storm Tides 7

2. A tug of war over the ocean 8

    2.1 Tides and gravity 8
    2.2 Newton's law of universal gravitation 9
    2.3 Tractive forces 9
    2.4 Static tide concepts - the equilibrium tide model 13
    2.4 Deductions from the equilibrium tide model 14
    2.5 Solar tides 16
    2.6 The spring-neap cycle 17
    2.7 Long-term variations in the tide 19
    2.8 Calculating the lunar tide producing force 22

3. Are tides waves? 25

    3.1 An encounter with wind waves 25
    3.2 A spectrum of waves superposed on the sea 26
    3.3 A look inside the waves 29
    3.4 Tides as long waves 32
    3.5 Ocean tides and Kelvin waves 33
    3.6 Progressive waves and standing waves 34
    3.7 Wave dynamics and wave energy 37
    3.8 Kelvin waves in ocean basins 38
    3.9 From rotary waves to sine waves 44

4. Tidal constituents: buildings blocks of the tide 46

    4.1 Harmonic tidal constituents 46
    4.2 The harmonic model of the tide 50
    4.3 Tidal types and the tidal form number 51
    4.4 Referencing time and height of tide 53
    4.5 Seasonal tides and shallow water tides 56
    4.6 Simulating tides with a series of cosine waves 58

Table of contents

5. Tidal datums: finding the apparent level of the sea — 60

    5.1 Definitions and use of a tidal datum — 60
    5.2 On the level? It's a geodetic datum — 61
    5.3 A tidal datum begins with a tide staff — 61
    5.4 And ends with a tidal bench mark — 63
    5.5 Tidal epochs: why 19 years? — 63
    5.6 Tidal datums in use around the world — 64
    5.7 Sea level trend and the tidal datum epoch — 64
    5.8 Transferring tidal datums — 67

6. Tides and currents in a large estuary - Chesapeake Bay — 73

    6.1 Tides inside Chesapeake Bay — 73
    6.2 Co-range and co-tidal charts — 74
    6.3 A simple shallow-water wave model — 75
    6.4 Tidal height and tidal current relationships — 83
    6.5 Rotary currents in lower Chesapeake Bay — 86
    6.6 Non-tidal currents — 88
    6.7 Tidal and non-tidal flushing in estuaries — 90

7. Analyzing tides and currents — 93

    7.1 What is harmonic analysis? — 93
    7.2 Harmonic analysis of water levels: *SIMPLY TIDES* — 96
    7.3 Some examples of water level analysis — 97
    7.4 Harmonic analysis of water currents: *SIMPLY CURRENTS* — 109
    7.5 Some examples of water current analysis — 110
    7.6 When tidal constants aren't — 120

8. Predicting tides and currents — 123

    8.1 The US NOS tidal prediction formula — 123
    8.2 Harmonic constants without equilibrium arguments — 125
    8.3 Tidal height predictions made simple: *SIMPLY TIDES* — 126
    8.4 Comparison of *SIMPLY TIDES* with US NOS tide predictions — 127
    8.5 Tidal current predictions made simple: *SIMPLY CURRENTS* — 131
    8.6 Comparison of *SIMPLY CURRENTS* predictions — 131
    8.7 The outlook for more and better current data — 132

9. Storm tide and storm surge — 134

    9.1 Beyond the astronomical tide — 134
    9.2 The US Saffir-Simpson hurricane scale — 135
    9.3 Some definitions — 135
    9.4 Storms of record in the United States — 136
    9.5 The importance of being referenced — 143
    9.6 Predicting storm surge and storm tide — 146

10. Computational methods: the matrix revisited   153

10.1 Advantages of MATLAB® matrix algebra   153
10.2 Harmonic analysis compared with spectral analysis   157
10.3 Harmonic analysis, method of least squares (HAMELS)   159
10.4 The MATLAB advantage   162
10.5 Serial date and time - Microsoft® Excel   162
10.6 Water level and water current analysis using HAMELS   163
10.6 Determining the principal axis for water currents   165

To Nancy, Katy and Zeke

Preface

As the worldwide community of coastal residents continues to grow, interest in the dynamic behavior of the 'coastal ocean' has likewise expanded. It is interest well placed in anticipating what the future will bring to this realm. We are all aware of the daily rise and fall of the tide, the flood and ebb of tidal currents, but are perhaps less conscious of the essential role they play toward keeping our seas and estuaries clean and healthy through their continual flushing action. We are keenly aware of the destructive storm tides that periodically strike and inundate our shores but know much less about seasonal change, secular sea level trends and the water level base from which future storm surges will arise. These and similar concerns are prompting renewed attention from many quarters. They come fortunately at a time when information technology has made the data needed for their study accessible worldwide.

This book offers an introduction to the subject of sea tides and tidal currents for readers who would like to understand these phenomena more clearly through a self-paced learning approach with a minimum of mathematical development. Its purpose is twofold: Firstly, it is intended as a resource for graduate and undergraduate students who want to pursue the subject beyond the brief introduction that usually accompanies a general course in marine science or oceanography. Secondly, it offers these students the tools needed to study actual tides and currents, to take them apart and put them together again in the form of predictions. Why tides and tidal currents are predictable years in advance, and how these predictions are made, are among the 'secrets' revealed in this book.

The origin of the tides and man's discovery of ways to predict them are two closely related subjects with a long history. Civilizations living by the sea in ancient times experienced the water's rise and fall and the current's flood and ebb as everyday events. They marked the passage of time by the phases of the moon and the sun's rise and set on the horizon. Human imagination ensured that many forms of the empirical 'art' of predicting tidal behavior were practiced well before theories explaining the tide's true origin rose to the cutting edge of science in the Newtonian age.

For those with a particular interest in tidal theory and its parallels with the development of physical science, David Cartwright's *Tides: A Scientific History* is a book well worth reading. Cartwright, an eminent scientist and pioneer in the field of oceanography, begins his history with a description of the earliest practitioners of the art, remarking on "the interest, experimental ability and persistence of people 2000 years ago to record the heights of the tide throughout the year, to note quite small changes, and to relate them to the ephemerides of the moon and sun". Cartwright then takes his readers forward in time through many discoveries, ending with today's supercomputer models of world ocean tides developed with the aid of satellite altimetry.

The present book focuses on present ways of making tidal predictions of the type coastal residents are most familiar with in everyday life: When will the next high or low tides occur and how high or low will they be? Familiar with the predictions yes, but with the way they are made? Not really. These 'ways' have, until now, been the somewhat cryptic and closely held property of a relative few behind closed doors. Technological advances made toward the end of the millennium have changed this picture considerably. Those with access to the World Wide Web and to certain popular software packages on their personal computer can now push the doors wide open.

To predict tides in the way described in this book, you first have to analyze some measured tides and derive what are known as the tidal harmonic constants. Once you have these numbers for a particular place, you can make tidal predictions there for any day, month, or year. Predictions, however, are only part of the fun. The method of tidal analysis described herein will also permit you to examine the surge, the 'meteorological tide' that accompanies coastal storms, after its separation from the astronomical or 'normal' tide. It's the surge that's usually meant when the media speak of tides so many feet above normal during an approaching storm.

The calculations required for tidal analysis are easily made, especially with the aid of matrix algebra. In fact, it may be closer to the truth to say that matrix algebra is essential unless your programming skills are exceptionally good. Taking advantage of its excellent matrix handling capabilities, the programs I describe in this book were written using the MATLAB® scientific programming language[1]. Most of the technical illustrations were constructed using MATLAB graphics routines. Other than the basic language, the two sets of programs offered herein require no MATLAB software accessories or 'toolboxes' but do require Microsoft® Excel for organization and storage of observational data. One set of programs handles the analysis and prediction of tides, the other the analysis and prediction of currents. Both employ a Graphical User Interface or GUI, making them very simple to use and obtain graphical output. Both are available for downloading at no cost from the MATLAB Central file exchange (http://www.mathworks.com/matlabcentral/fileexchange/).

The mathematical technique most commonly employed in analyzing tides is called *harmonic analysis by the method of least squares* or **HAMELS**, the acronym I use in describing this approach. Technical discussions concerning HAMELS and its implementation with the aid of matrix algebra are presented in Chapter 10 at the end of the book. Although the technical discussions could stand alone, they would not read very well without several introductory chapters containing essential information about tidal behavior - the origin of tides and their method of propagation as long waves in systems ranging from large ocean basins to coastal estuaries. The role of tidal datums and sea level change, two related and highly relevant topics, also require an introduction before tide prediction can take on any practical value. I chose to begin with an introduction to the Tidewater region of Maryland and Virginia – a region worth examining because it provides many excellent examples of why we continue to study tidal behavior in an estuarine system with only modest tides in terms of their daily limits of rise and fall, and why that behavior has remained important from historical to modern times. Through this introduction, I hope I will have given you added reason to want to know more about tides and currents in other regions of the world.

My own interest in tides came after growing up near a lake in central Texas. There are no tides in Texas lakes but an uncle who served as a naval officer in World War II must have sensed my nascent interest when I was a boy. He arranged a trip to the city of Galveston on the Gulf of Mexico. Well do I remember that day. Long before we reached the bridge leading to Galveston Island, the Gulf was heralded by its salt scent. I quickly discovered that the Gulf was no mere lake. In fact it was huge and I couldn't see beyond its horizons. Then there were the waves – large breakers – that made a

---

[1] MATLAB is a registered trademark of the MathWorks, Inc.

hissing, thundering sound as they spilled and collapsed onto the beach, one after the other. As each wave struck the sloping beach face it surged forward, stopped, and then rushed seaward again as backwash accompanied by the unique, tinkling sound of colliding sand grains and tumbling shells. Near the end of its run at the foot of the beach, the powerful backwash dislodged a horde of startled creatures – mole crabs and tiny coquina clams – all digging frantically to escape under the sand before the next wave hit them. Fresh rows of gulfweed, a free-floating algae supported at sea by berrylike air sacs, lay several feet beyond the limit of the wave's up rush on dry beach.

I was told all this meant we were visiting at low tide. That led me to ask my uncle: 'what *is* the tide?' His answer was spellbinding. For the first time I heard about the rhythm of the tide, the rise and fall of the water accompanying its ceaseless excursions back and forth across the shore, the role of the moon and sun – all positively reinforced by the natural ambiance of a beautiful, fair-weather day. But the sea had a dark side, my uncle explained. The early Spaniards had called Galveston the Isla de Malhado (Isle of Misfortune). On September 8, 1900, Galveston lived up to that name as one of the deadliest hurricanes in history produced huge waves and a massive storm tide that swept over the entire island, drowning more than 6000 of its inhabitants with little warning. I still try to imagine today what it must have been like for those ill-fated souls, caught in the dead of night and surrounded by rising water. Unlike the sand-dwelling mole crabs, they could not escape the merciless wind and waves.

I would like to express my appreciation to the National Ocean Service, National Oceanographic and Atmospheric Administration, U.S. Department of Commerce, for making their data freely available via the web. In the United Kingdom, the National Tidal & Sea Level Facility, British Oceanographic Data Centre, and the Proudman Oceanographic Laboratory are likewise thanked for granting web access to their tide and current archives. The Comune di Venezia and the Venice Tide Forecasting Centre provided other data for Venice, Italy. I am indebted to my colleagues John Brubaker, Carl Friedrichs, Malcolm Green, Albert Kuo, and Art Trembanis for reading parts of the initial manuscript and for offering their valuable comments. I owe special thanks to my friend and mentor, Bob Byrne, who encouraged me to write this book and offered his guidance throughout. I am grateful to Mr. Ellis Horwood, my publisher, for his keen interest in the project and for wisely encouraging a broad scope with worldwide examples. Lastly, I thank my wife Molly for proofreading the chapters and for bearing with me as the task rolled on.

John Boon

Gloucester Point
August 2004

1

Tidewater

1.1 TIDEWATER GEOGRAPHY

Any region where water is activated by the tide could be called '*tidewater*'. Not every region, however, employs the term both as a geographical descriptor and as a basis for its historical and cultural identity. Within the eastern United States, only Virginia and Maryland have used it in exactly this way since their beginning days as English colonies. Geographically, Tidewater Virginia and Tidewater Maryland together form a triangle whose base extends from the Chesapeake Bay entrance westward through Hampton Roads and the lowlands south of the James River to Petersburg in Fig. 1.1.

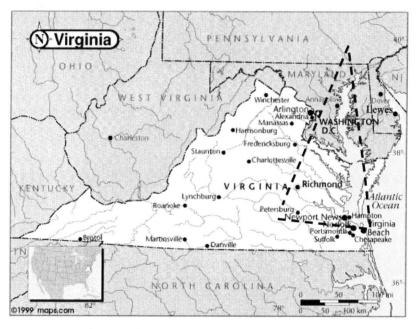

Fig. 1.1. Map showing the Tidewater region of Maryland and Virginia (dashed triangle).

The west side of the triangle shown above continues northward from Petersburg along the *fall line* (**head of tide**) connecting the river front cities of Richmond, Fredericksburg, Washington, and Baltimore. The east side descends from the northern tip of Chesapeake Bay southward along the Eastern Shore of Maryland and Virginia back to the Bay entrance. The eastern boundary is broadened a bit to include both the western (bay side) and eastern (ocean side) shores of the Maryland-Virginia Peninsula.

1

Within this combined region centered about Chesapeake Bay lie more than ten thousand miles of tidally influenced shoreline.

If you ask a marine scientist or an oceanographer to describe the tide in Chesapeake Bay, they will have this reply: **microtidal**. To get the meaning of this term we have to define **tidal range**. Simply put, tidal range is the vertical distance covered by rising and falling water levels due to the action of the **tide**. A more specific definition will be given later, but for now just consider that microtidal refers to an average tidal range throughout the region of less than 2 meters (6.6 feet). In fact, tidal range averages less than half this amount throughout most of Chesapeake Bay and its river tributaries.

Compared to **macrotidal** regions, such as the Bay of Fundy in Nova Scotia where tidal ranges exceeding 12 meters (40 feet) are not uncommon, Chesapeake Bay tides would seem inconsequential. However, that is far from being true. Not only are tides important in tidewater Virginia and Maryland but **tidal currents** are especially so. The periodic rise and fall of the tide in coastal embayments like the Chesapeake Bay and its tributaries means that **flood** (landward moving) and **ebb** (seaward moving) currents must exist to transport ocean water in and out of these systems during every tidal cycle of approximately half a day. When the horizontal area covered by tidal water is large, as it is in Chesapeake Bay, one of the world's largest estuaries, the volume of water in transit (area times tidal range) is also large so that strong currents can be expected in narrow entrances. However, tides come from the ocean. Building a dam across the entrance between Cape Henry and Cape Charles, for example, would ensure the end of measurable tides - and tidal currents - throughout the Chesapeake Bay. This unlikely prospect is likely to remain beyond anyone's serious consideration for the foreseeable future, although the same has not been true for coastal seas and embayments elsewhere such as the Ijsselmeer and the Scheldt region in the Netherlands.

Tides and tidal currents clearly influenced life in colonial Virginia at a time when its commerce relied on ships and ships relied on sail power aided by the tide. Plantations were sited along the thousands of miles of winding shoreline protruding inland along the major river estuaries, the James, York, Rappahannock, and Potomac Rivers, plus many smaller tributaries, all part of the western tidewater. Plantation houses always faced the river and there were no roads leading anywhere else. Everything came and went by water, often over a considerable distance.

Sailing with or against the tide, for example, might easily bring a ship from the Bay entrance to the mouth of the Potomac River, but from there more than a hundred miles of confined and difficult sailing would remain to get to the innermost ports of that river. The passage was made easier by anchoring the vessel during ebb and sailing or towing it further upstream during flood. The reverse would apply when leaving port. It was no accident that major cities arose at the ends of these lines of transport - the limit of tide at the head of each tidal waterway.

The benefit of a microtidal environment in this region is thus apparent and clearly fortunate considering its generally low-lying position and gentle relief. A tidal range of 40 feet between Norfolk and Richmond would be unimaginable in this setting.

1.2 EARLY USE OF TIDAL POWER

Even without a 40-foot tide range, tides were exploited for more than navigational commerce. By the mid-eighteenth century *tide mills* were used in a number of locations, including Gloucester and Matthews Counties in Virginia, for grinding grain into flour or meal. Little remains of these mills today but you can still see the large millstone from one of them near the Tide Mill store in Hayes, Virginia, at the head of a small creek entering the York River. It is said that this mill ground grain for Washington's forces during the siege of Yorktown in 1781. Although there are no working tide mills in the United States today, there is at least one still in operation in the United Kingdom; More than 900 years old, the Eling Tide Mill in Southampton Water on the south coast of England is perhaps the earliest source of this technology that was later used in the British colonies in America.

It should be pointed out that there were two kinds of mills in the lower tidewater region: both of them located a considerable distance from the fall line and swift-flowing rivers. One type was usually situated at the mouth of a millpond formed by damming an inland creek in areas with moderate relief upstream (a winding escarpment along one of the ancient bay shorelines usually provided the relief). This type of mill derived its mechanical power from a large water wheel driven by water flowing out of a small sluice gate or pipe over the top of the wheel. The pond was 'recharged' by groundwater and runoff.

The other type was a mill placed along a tidal creek partially dammed so that all of the flood and ebb flow passed under the water wheel through a narrow canal. This would drive the water wheel first in one direction and then the other as the tide reversed. For this to work, the diameter of the wheel needed to be at least twice as large as the tidal range so that only its lower half was immersed.

Unlike the 'pond' mill, a tide mill could operate only a part of each day, being shut down entirely during the *slack water* period that always occurred in the interval between a flood or ebb and the following tide. The average tidal power generated by one of these mills varied roughly as the square of the tidal range, meaning that if the range could be doubled, then the tidal power would go up by a factor of four. Although the tidal range in the lower York River averages 2.4 feet (0.73 meters) and increases by about 15-20% during the so-called *spring-neap cycle* (see Chapter 2), this was a bit less than needed to make tidal power for milling much of a success in the region.

Tide mills were more common in the New England region north of Cape Cod where tidal ranges of 9 feet (2.7 meters) or more occur.

1.3 THE INTERTIDAL ZONE

Just as tidal range combined with water surface area determines the magnitude of tidal currents in bays connected to the ocean, the combination of tidal range and bottom slope at the shoreline determines the **intertidal zone** width. This is the zone at the land's margin that is covered and uncovered by water during an average tidal cycle. On a day when the tidal range is 1.2 meters (4 feet) at Wachapreague, Virginia, land that rises 1 meter for every 10 meters (1:10 slope) measured horizontally across the shoreline would have an intertidal zone 12 meters (40 feet) in width. If the slope were 1:20 instead of 1:10, then the intertidal zone width would be 24 meters (80 feet).

The width of the intertidal zone varies from place to place and from one day to the next but over the long term, it delineates a special type of habitat that includes most of the **wetlands** of Virginia and Maryland. At the low end of that habitat we find (or used to find) shellfish such as oysters. There are many reasons for the decline of the once enormous oyster populations but one view in particular holds that these communal organisms were most abundant when living in shallow banks called oyster reefs. In appearance, these are like curving river sand bars except for being composed of masses of oysters, live ones successively attaching to, and living upon, their dead predecessors below. Even away from the shoreline, these bars were part of the intertidal zone and the animals living on them would be exposed to the air for at least a few hours each day.

Covering and uncovering by water not only tended to wash away sediment and other debris but may have thwarted marine predators such as the oyster drill. As the name implies, an oyster drill is a small mollusk that kills oysters by drilling a hole in them. Unlike the oyster, it is free roaming and unattached. However, it must do its work underwater and thus had no choice but to give the reef oyster a reprieve at low tide. With the return of higher water levels, the drill faced the improbable task of finding the same hole to resume any uncompleted work. Human predators working with tongs gave no such reprieve and artificial oyster reefs, man made banks constructed of fossil shells, are virtually the only ones in existence today and these owe thanks to local restoration efforts.

1.4 TIDAL WETLANDS

Without doubt one of the most remarkable characteristics of the intertidal zone is the unique habitat it creates for plants as well as animals. Perhaps the most visible sign and logo for the tidewater region is its extensive marshlands composed of salt and freshwater plant species. Marshes by their very nature are creations of the tide. The fine sediment or mud forming the base of the marsh must first be brought to its point of deposition by tidal currents. But the intertidal zone, accessible to any seed or rhizome brought by wind or water, poses a tough question to newcomers: "Can you sustain yourself in this muddy soil for so-many-hours under water and so-many-hours above water each day?"

In areas under the ocean's influence, saltwater tolerance is added to that test. In Virginia and Maryland, few species among either land or aquatic plants can tolerate a daily dunking in and out of saltwater. Those that can tend to group themselves into banded zones characterized by a certain frequency of immersion (daily, once a month, once a year, etc.). For example, the dominant plant found in the extensive salt marshes on Virginia's eastern shore is the cord grass, Spartina alterniflora. Due to its unique cell structure and physiology, it can tolerate immersion in high salinity water for up to 17 hours a day although it isn't 'obligated' to do so. That means it can also grow in places where the immersion rate isn't so high but there, of course, it will face increasing competition from other species. In fact, a related species, Spartina patens, does just that and situates itself comfortably in the zone lying just above S. alterniflora. Several other wetlands plants can insert themselves in additional vegetative bands until the upper limit of the intertidal zone is reached, a zone that experiences immersion only once or twice a year at most. At that level a gradation into strictly land plants takes place. The interesting thing about this zonation in saltwater marshes is that each species tends to have a fairly sharp lower limit but a less well-defined upper limit in terms of the elevation where it grows. It's as if the plants are saying "We can take saltwater immersion at just this rate and no more!" It should not be surprising that this is not the case in tidal fresh water marshes where diversity is high and many species tend to crowd together within the intertidal zone.

Tidal marshes are more than just interesting and aesthetically pleasing creations of the tide. They form a valuable protective buffer between land and sea. Dense stands of salt marsh cord grass, well anchored by thick, peat-generating roots, can provide a formidable barrier against destructive wave action on an otherwise unprotected shoreline. Protection comes in two ways. First, the tough, spongy peat layer generated by the roots (rhizomes) of S. alterniflora can take quite a pounding from breaking waves. When failure does occur, it is mainly through scouring underneath the peat, especially in sandy soils along riverbanks. Secondly, as we will see in Chapter 3, waves move forward (propagate) because of the energy residing in water particles that are engaged in a circular motion instead of being at rest. The heavy stalks and stems of the cord grass tend to interrupt that motion through frictional resistance, progressively bleeding away the energy of the waves as they travel across the marsh at high tide. Even a *fringing marsh* only a dozen feet in width, but possessing a good peat layer, can compare favorably with a stone or timber bulkhead costing more than a thousand dollars a foot. The practical value of marshes has been further enhanced through recognition of their role as filters of pollutants entering waterways through land runoff, in addition to their importance as a nursery ground and sanctuary for juvenile fishes and crabs.

1.5 TIDAL CIRCULATION AND TIDAL MIXING IN ESTUARIES

Besides the visible attributes of the tide that local residents witness every day, unseen action is constantly present, whether we know it or not, in the form of **tidal circulation** in Chesapeake Bay and its tributaries. This circulation occurs because the Bay and its tidally-influenced tributary systems are examples of an **estuary**, a body of water in which sea water from the ocean is measurably diluted by fresh water discharged from inland rivers, creeks and underground flow. Much of the fresh water enters at the heads of the major rivers on the west side of the Bay: the James, York, Rappahannock, and Potomac Rivers. These rivers are a bit unusual in that they are tidal and behave as *sub-estuaries*. That is, they behave that way until the fall line previously mentioned comes into view as one travels inland. Just beyond the limit of tide, each river reverts to its conventional mode marked by a relatively constant water flow in only one direction (downhill and seaward), spilling over rocks and producing waterfalls. With the exception of the York River, a relatively short sub-estuary that divides into smaller tributaries well below the fall line, these rivers continue inland under the same name. A fifth major river, the Susquehanna River, enters the northern extremity of Chesapeake Bay and contributes almost a third of the total freshwater inflow received by the Bay during the spring runoff period. Lesser inflows occur from countless small creeks and groundwater sources scattered along the way toward the Bay mouth.

Once freshwater enters an estuary, there is the question "where does it go?" Clearly it will try to keep on moving seaward but eventually it must meet, and form some kind of interface with, the salt water entering the estuary from the ocean since there are no obvious barriers, like a dam, keeping the two water types apart. A better question might then be "what keeps saltwater and freshwater apart?" The answer is gravity and a difference in density. The salt content (*salinity*) of seawater causes it to have a higher density than freshwater, ignoring other differences such as water temperature (cold water is denser than warm water of the same salinity). Carefully placing freshwater over saltwater in a jar results in a two-layer system with a lighter freshwater layer overlying a denser (heavier) saltwater layer - until someone shakes the jar and mixes them together. Using the jar model, it's not only possible to envision a two-layered estuary with saltwater underlying freshwater, but to expect the layering (**stratification**) to remain in place until it is weakened or eliminated through mixing. In the real world we can expect total **de-stratification** to take some effort. A jar of water, after all, is miniscule compared to a large river. Jumping from micro- to macro-scale means that the forces at hand and the masses of water being moved out of their comfortable gravitational equilibrium will undergo a similar change of scale. Clearly considerable amounts of energy are required to 'stir' an estuary! Much of that energy ultimately comes from the tide, or more specifically the tidal circulation, the masses of water put in motion by tidal forces.

Other forces include the wind acting on the water's surface. Wind stress puts the surface layer of water in motion, and that layer tends to drag along the next layer below it (through frictional coupling), transmitting energy downward, layer-by-layer. Tidal currents flowing over the bottom experience essentially the same process in reverse: frictional coupling with the bottom slows down the bottom water layer, which in turn slows down the layer above it and so on until a certain velocity gradient is set up in the vertical direction. If velocities are high and gradients steep, a strong shearing action takes place between layers, an action that tends to tear them apart. A diver suddenly swimming into an unseen gradient of this kind would call it turbulence. Ultimately turbulence accomplishes mixing. How much force is needed to generate enough turbulence to complete the mixing? That depends on how large the difference in density between layers happens to be, which in turn depends on the relative amounts of freshwater from rivers and saltwater from the ocean coming together in the estuary. As a rough guide, marine scientists compare the daily volume of freshwater inflow to the **tidal prism** (tidal range times the surface area of the estuary), using the ratio as an index for the expected level of stratification in an estuary.

Why is this important? One reason is the effect of stratification on tidal transport. Most commonly we view tidal currents as streams of water moving horizontally from place to place like a conveyor belt, carrying along anything that may be dissolved or suspended in the water itself (oxygen, nutrients, sediments, organisms). However, a characteristic of partially mixed estuaries is that the net movement of water over a tidal cycle in upper layers is usually in the seaward direction while lower layers tend to experience net motion in the landward direction (see Chapter 6). Juvenile fish and shellfish, as well as their larval counterparts, use this rather convenient transportation system to get where they need to be in the estuarine system at critical stages in their life cycle. Almost all such organisms have at least some swimming ability that enables them to move up or down in order to get into the right layer, a layer that will move them over far greater distances than they normally can swim, in the direction required. Non-living materials (e.g., nutrients) lack this ability and are dependent on the degree of mixing in the estuary for their dispersal upward or downward in the water column.

1.6 EFFECTS OF EXTREME STRATIFICATION: HYPOXIA AND ANOXIA

In places where density stratification is exceptionally strong, vertical transport is likely to become very difficult. In this situation, very little exchange exists between one conveyor belt and the next because large density differences in the corresponding water layers allow very little mixing. Without mixing, the layers flow smoothly past one another and act like a brake on turbulent exchanges upward or downward. The only materials likely to get past this barrier, in fact, are bits of organic debris consisting of larval remains and other forms of solid matter that gravity selects for a slow, one-way trip to the bottom. Lower layers, especially deep bottom layers, may then become languid semi-stagnant collectors of fallout from above. While this 'capping off' by upper layers persists, bottom waters slowly lose dissolved oxygen through biogeochemical processes (respiration, oxidation of organic matter in water and bottom sediment) and have no means of getting more.

The result is a condition known as **hypoxia** (deficiency of oxygen) or even worse, **anoxia** (no oxygen) that is lethal to free-swimming as well as *benthic* (bottom-dwelling) organisms. Periods of high freshwater runoff, as happens in spring, followed by long days of hot weather with no wind, as often happens in summer, provide the ideal combination of events to set this up. Anoxia has been documented in Chesapeake Bay since the 1930's and is particularly evident in river estuary basins where summer stratification is strong and persistent over the deepest layers. All of the major rivers on the western side of the Bay have deep basins near their mouths.

1.7 STORM TIDES

Wind, or wind stress on the water's surface, influences the dynamics of tidal estuaries. No wind blows for very long over an open stretch of water without causing **wind waves** to form. Strong winds not only generate large waves, they cause more of the energy in the atmosphere to find its way into the hydrosphere. Some of this energy actually couples with the water to set up wind-driven currents. Most of it remains fixed in the waves themselves whose waveforms move along in the direction of the wind at a rate considerably faster than the current. Depending on the characteristics of the wind waves being generated (**forced waves**), or those moving through an area after being generated (**free waves**), a sense of their energy may or may not be apparent to a local observer riding through them in, say, a fishing boat.

But the energy brought by storm waves traveling over perhaps hundreds of miles is all too apparent when, in the space of a few yards, it is suddenly expended through the process of wave breaking at the shoreline. This, too, is a common feature of the tidewater region. During major storms ranging from 'northeasters' to hurricanes, a different type of wave is generated called a **storm surge**. Large and very damaging wind waves often ride on the back of a major storm surge that raises the water level substantially and thus increases the distance inland where the waves can attack the shore. The way in which the tide combines with the storm surge has much to do with the characteristics of the **storm tide** that results.

This broad sampling of the myriad features that interact with the tide in often subtle but sometimes striking ways underscores the importance of the microtidal environments found in Chesapeake Bay and many similar systems around the world. They are indeed interesting and worth a closer look.

2

A tug of war over the ocean

Ocean tides are the result of a never-ending tug of war between the earth and the moon, a tireless contest going on for millions of years in which each player pulls not on a rope but the ocean itself. A third player is the sun, a body that also maintains a grip, although a lesser one, on this watery prize. The earth is easily winning the contest and has the distinction of being the only planet that we know of with a surface covered mostly by water: water that it has been able to keep but not without some sign of a struggle. Evidence of the struggle appears in the rise and fall of the tide in the world's seas and oceans, not at random but in lock step with the motions of the earth, moon, and sun. The invisible hand responsible for this state of affairs is the mysterious force known as gravity. Let's begin by exploring some of the history behind tides and gravity.

2.1 TIDES AND GRAVITY
Ocean tides have been observed at the edge of every inhabited continent for a very long time. Gravity, while always a part of the human experience, was nevertheless slow in developing as an explainable concept. It was equally long before someone finally made a firm connection between the two. In the seventeenth century, the renowned Italian physicist and astronomer, Galileo, came close to being that someone.

Galileo was at once a scientist, inventor and experimentalist, the world's best in his day and time. After inventing the astronomical telescope, he was the first to see the mountains and valleys of the moon. He saw that Jupiter also has moons and noted that they orbited that planet with the precise timing of an extremely fine watch (*tic*, another revolution, *tic*, another revolution,). Regarding gravity, it was Galileo who first demonstrated through experiment that the rate of increase in the speed of falling objects (cannonballs dropped, according to legend, from Pisa's Leaning Tower) was a constant, no matter how big or how small the objects were, after making an allowance for air friction.

This constant, about 9.8 m/sec^2 (32 ft/sec^2), is called the acceleration of gravity and scientists everywhere describe it by the lower case letter g. It's the same g that jet pilots refer to when they make a tight turn and say "I pulled two g's that time!" meaning that the pilot experienced two times the normal force of gravity pulling him or her down in their seat during the turn. The unit of measurement for gravity is the *gal*, named after Galileo. Regarding tides, Galileo offered his own observations of tides in the Mediterranean Sea as one proof that the earth did not sit motionless at the center of the universe, as the Greek philosopher Aristotle had claimed, but moved spinning in an orbit around the sun, as the Polish astronomer Copernicus proposed. Galileo noticed that water placed in barrels aboard a ship sloshed up and down against the sides of the barrel as the ship moved. Wouldn't the water in the sea do the same thing as the earth moved? Though he was right in siding with Copernicus, an action that almost cost him his life at the hands of the Inquisition in Rome, Galileo's theory on the origin of the tide was fatally flawed because it overlooked any reference to gravity beyond that of the earth.

Isaac Newton, the famed English physicist and mathematician, was born in 1642, the year Galileo died. Newton gave the world its first clear set of rules, or scientific laws, governing the motion of objects and defining how gravity works (*what* gravity is remains something of a mystery even today). He invented a new branch of mathematics, the calculus, to aid in developing his theories and to allow their rigorous comparison with observations. Along with his laws of motion, Newton presented his own theory on the origin of tides, a theory that is still in use today. Before discussing the connection between gravity and ocean tides that Newton made, we first need to see how gravity works over the large distances between the earth, moon and sun. Newton's remarkable discovery was that, through gravity, each of these bodies attracts not only the other two but also attracts every other body in the universe, all at the same time.

2.2 NEWTON'S LAW OF UNIVERSAL GRAVITATION
Newton's law of universal gravitation states that the force of mutual attraction between any two bodies varies directly as the product of their masses and inversely as the square of the distance between them. A simple equation expresses this relationship:

$$F = G\frac{m_1 m_2}{R^2}$$

where the letter F represents the force of mutual gravitational attraction, m_1 is the mass of the first body, m_2 is the mass of the second body, and R is the distance between the bodies (the distance between their centers, or more correctly, their centers of gravity). G is a proportional constant known as Newton's gravitational constant, not to be confused with g, the acceleration of earth's gravity. If m_1 is the mass of the earth and m_2 is the mass of a person standing on the earth (a body after all), then R is the radius of the earth and F is the force of mutual attraction or simply that person's weight. If instead we take m_1 to be the mass of the moon, then F represents the mutual attraction between the person and the moon. This is obviously a much smaller force because the moon's mass is much less than the earth's and the product $m_1 m_2$ is now a smaller number in the above equation. Also it is being divided by a larger number, R^2, where R is now the distance between the moon and the person standing on the earth. It happens that the earth's mass is about 81.5 times that of the moon and our distance from the moon is about 60.3 times the radius of the earth. Doing the math (see Sec. 2.9 CALCULATING THE LUNAR TIDE PRODUCING FORCE) shows that whatever the person's weight may be, the moon's attractive force toward him or her is only 0.0000034 times as great when averaged over the earth. The same would be true if we used a barrel of water, in place of a person, as the example. Galileo can be easily forgiven for not devising an experiment that would detect a force so minute.

2.3 TRACTIVE FORCES
How can a force so small still have the ability to produce the tide? We have to do a bit of digging to answer that question. To start with, we know that the masses of water put

in motion as tides rise and fall all over the world are many times larger than that of a single barrel. But whatever the quantity of water involved, when we compare the gravitational forces acting on it, the moon's versus the earth's, the ratio is still the same: 0.0000034. Here's where the difference comes in.

As everyone knows, the moon slowly orbits the earth. More precisely, the earth and the moon together orbit a common center of gravity, a point called the **barycenter** that lies just inside the earth. In doing so, a *centrifugal* force is set up at the center of each body that exactly counterbalances the gravitational force tending to pull them together. We experience centrifugal force when we swing a weight attached to a string over our heads. Cancelling the gravitational force between moon and earth would be the equivalent of breaking the string; Newton's law assures that we can count on that not happening. Overall, the moon's gravitational force will exactly match the opposing centrifugal force on the earth in a state of equilibrium that will continue indefinitely.

Looking at Fig. 2.1, we see a diagram of the earth in a cross-section through its center. The moon is in the direction of the line leading off to the right. Assuming the earth's radius to be 1 inch in the drawing, the distance to the moon would be about 60 inches or five feet off the page. Gravitational attraction between the moon and the earth, F_M, and centrifugal force on the earth, F_C, are represented by two arrows of equal length shown at the center of the earth, pointing in opposite directions. This signifies not only that the forces on the earth as a whole are in balance but it also means that a separate object at the center of the earth would also experience a *net force* of zero for F_M and F_C calculated using the mass of the object in place of the earth's mass. The question is, does the sum of F_M and F_C remain zero at other points on the earth? The answer is no.

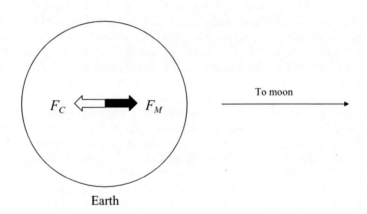

Fig. 2.1. Balance between centrifugal force, F_C, and the moon's gravitational force, F_M, acting in opposition at the earth's center.

Referring again to Newton's law of universal gravitation and the equation we used to represent it, we notice once again the R^2 term in the denominator. If we move away from the earth's center to the point on the earth's surface nearest the moon (the 3 o'clock position in Fig. 2.2 below), we decrease R by one earth radius and increase the moon's gravitational force per unit mass, F_M, from 0.0000034g's to about 0.0000035g's using the jargon of our jet pilot. Moving to the point on the earth's surface farthest away from the moon (9 o'clock position in Fig. 2.2), R would increase by one earth radius and F_M would decrease to about 0.0000033g's, a change of about 3%. The centrifugal force at the center and at every other point on the earth remains the same as the earth orbits the barycenter without rotation; i.e., constant in magnitude and directed away from the moon.

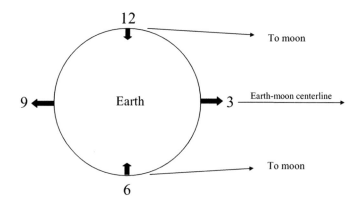

Fig. 2.2. Differential forces acting at the earth's surface.

As small as these *differential* forces are, they can and do have a significant effect on the huge volumes of water in the ocean that, unlike the solid earth, are quite free to move – in certain directions. To an observer, the two surface points at 3 o'clock and 9 o'clock lie directly *under* the moon as shown in Fig. 2.2. The differential forces at each of these points would therefore act to lift the water straight up relative to the earth's surface in direct opposition to earth's gravitational force, F_E, which weighs in at a full 1 g. This is like removing a mouse sitting on the back of an elephant – the elephant isn't going to feel any different after the mouse is gone. Likewise, a differential force exists at the 6 o'clock and 12 o'clock positions on the earth's surface as shown in Fig. 2.2. Note that this force, however, is directed inward toward the center of the earth. It arises not because F_M differs in magnitude from the attractive force at the earth's center (the distance to the moon is about the same) but because it differs in direction. Lines leading toward the moon from these points, along which F_M must act, are bent slightly inward toward the earth-moon centerline.

As a result the centrifugal force, which remains parallel to the centerline, is not completely balanced by the gravitational force at this point. The force 'left over' (the vector resultant) is a differential force directed toward the center of the earth. So we have done little more than add another mouse to the back of the elephant. Or make that a herd of mice. Rather than acting at just two points as shown in Fig. 2.2, the inward differential force acts everywhere along a great circle passing through the 12 and 6 o'clock positions in a plane perpendicular to the page. All of these inward forces have little effect on the earth's gravitational force, F_E, and, like the forces at 3 o'clock and 9 o'clock, have no tide-producing role. Clearly the earth wins the tug of war at all of these points.

But we are not done yet. Take another look at Fig. 2.2 and imagine what happens at points on the earth in between the ones just discussed. In passing from the 12 o'clock to the 3 o'clock position, the differential force has to progress from one directed toward the earth's center to one directed fully away from it. Fig. 2.3 shows how the differential forces at 1 o'clock and 2 o'clock change their direction so that they yield a horizontal component, a part of their force that acts parallel to the surface of the earth instead of perpendicular to it. This part is called the **tractive force**.

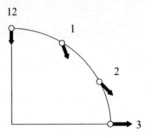

Fig. 2.3. Change in direction of differential forces.

A tractive force, 'traction', always acts parallel to a surface, a road for example, and is what sets a car in motion as its wheels turn. Tides are now possible because of one important fact: Earth's gravity does not oppose this force. Just as the mouse we spoke of is free to run across the elephant's back, ocean water is free to move across the earth's surface in response to the tractive forces. These forces are shown in Fig. 2.4.

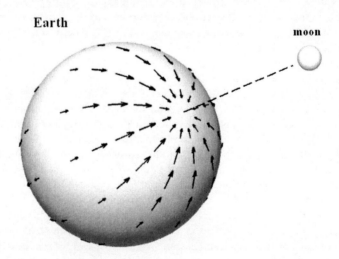

Fig. 2.4. A net of lunar tractive forces cast over the surface of the earth.

Like something out of science fiction, the moon's tractive forces spread over the earth's surface to form a gravitational force field, the net-like structure in Fig. 2.4. The length of each arrow in this net indicates the relative strength of the tractive force and the arrow's direction shows which way the force is acting. Two things about the net are particularly useful in understanding how tides originate. First is the tendency of the forces to converge on two points, one on the side of the earth nearest to the moon and one on the side farthest away . The second is the constant change in the tractive force sensed at points on the earth as it turns inside the net.

2.4 STATIC TIDE CONCEPTS – THE EQUILIBRIUM TIDE MODEL

If we imagined a frictionless earth completely covered by a frictionless layer of water free to respond instantly to the tractive forces acting upon it, the water would deform and eventually reach an equilibrium state. This state often appears in textbooks in the form of two 'tidal bulges' arranged on either side of the earth in line with the moon. An exaggerated cross-sectional view of a pair of idealized tidal bulges appears in Fig. 2.5.

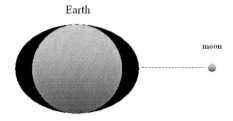

Fig. 2.5. Earth covered by two imaginary tidal bulges in line with the moon.

Here's a brief explanation of the 'static' tide concept. If we assume the outer shell of water covering the earth responds instantly to the lunar tractive forces depicted in Fig. 2.4, it will deform from an otherwise spherical shape. Once this happens, a component of earth's gravity will arise offsetting the deformation until a static equilibrium is reached in the form of a prolate spheroid whose major axis is aligned with the earth-moon centerline. Since we're ignoring friction, the earth is meanwhile free to rotate on its axis without disturbing the prolate spheroid and its peaks (the imagined tidal bulges) surrounding it. In this way, it is possible to explain how a **semidiurnal tide** with two high waters and two low waters occurs in most places on the earth each **lunar day**. Now let's turn to an example to see how that might occur.

Fig. 2.6 shows an imaginary person, Joe, in position at one of the northern latitudes on a rotating earth. As the earth rotates on its axis, Joe will move with it. He will encounter two high tides, corresponding to the two bulges, and two low tides, corresponding to the water-deficient points halfway between the bulges, each time the earth completes one rotation with respect to the moon. One earth rotation with respect to the sun marks the usual 24-hour solar day. However, the moon travels in an easterly orbit around the earth, moving in the same direction as the earth as it turns on its axis. As a result, a rotation with respect to the moon takes a little while longer - about 50 minutes longer - so that a *lunar day* averages about 24 hours and 50 minutes. During this time, Joe will encounter a high water just as the moon passes overhead during the upper transit (across his local meridian[1]) and again 12 hours and 25 minutes later as the moon completes the lower transit (passes the opposite meridian on the other side of the earth).

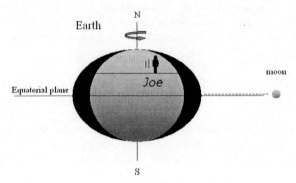

Fig. 2.6. Joe on a rotating earth with the moon over the equator

Like Joe, the two tidal bulges traveling all the way around the earth in Fig. 2.6 are not real. But they don't have to be. In many situations it is possible to derive insight by first making a simple, if not entirely correct, model of how things work. Isaac Newton, while realizing that the earth's surface is not entirely covered by water, nevertheless found the above concept very useful. Newton advanced the **equilibrium tide theory** by assuming a prolate spheroid would result from the gravitational forces he described. The theoretical range of tide (vertical distance between high and low water) could also be computed from his theory. Interestingly, Newton did not envision tractive forces nor did he predict the time of high water to be coincident with lunar passage above the local meridian. Others added these refinements at a later time. The sun as well as the moon was deemed a part of his equilibrium theory and he was successful in explaining phenomena long observed among real tides as noted below. Indeed the equilibrium theory has proven particularly useful as a standard for comparison with real tides, noting that a significant time difference is usually found between actual high tide and the equilibrium tide marked by moon's crossing of the local meridian. The average difference, the **lunitidal interval**, is a constant at each place where measurements have been made to determine it. Likewise, the mean range of tide at a place is found to be approximately a constant forming a fixed ratio with the equilibrium range.

2.5 DEDUCTIONS FROM THE EQUILIBIUM TIDE MODEL
We can learn a number of things about tidal behavior from this model. First we'll equip Joe with a measuring stick **(tide staff)** and we'll have him nail it to a piling on someone's ocean pier. Fig. 2.6 suggests that while the moon lies within the **equatorial plane**, the alignment of the bulges will be such that Joe will measure two high waters of equal height, and two low waters of equal height, on his staff each day. The moon meanwhile continues to orbit the earth in a plane that is inclined to the earth's equatorial plane. As a result, it will rise above, fall below, and then rise again through the equatorial plane to finish a single orbit as shown in Fig. 2.7.

---

[1] A meridian is a line of constant longitude, half of a great circle, on the earth's surface.

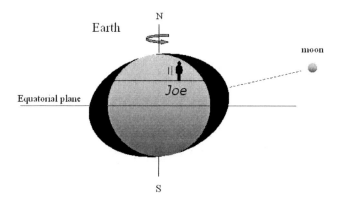

Fig. 2.7. Joe observing unequal tides with the moon north of the equator.

One of three 'monthly' cycles associated with lunar motion, this is called the **lunar declinational (tropic-equatorial) cycle**; it requires 27.32 days to complete. Assuming we start the cycle with the moon passing through the equatorial plane, as in Fig. 2.6, the highs are of equal height and so are the lows - We call these **equatorial tides**. But when the moon is at maximum declination north or south of the equatorial plane, as in Fig. 2.7, the tidal bulges show a corresponding shift. You can see that, as Joe moves with the earth, he will again measure two high tides in the course of a lunar day but one of them will be significantly higher than the other. These are called **tropic tides**. Tropic tides can have a pronounced **diurnal inequality** in the heights of either the highs or the lows (or both) depending on the region. In the extreme, some regions may experience a **diurnal tide** with only one high and one low per lunar day (see Chapter 3, Sec. 3.8). But whatever their form, the diurnal inequalities reach their peak at maximum lunar declination, north or south of the equator, and disappear during equatorial tides, just as the model predicts.

Besides circling the earth in an orbit inclined to the equatorial plane, the moon has one other attribute that affects the tide: its orbit around the earth describes an *ellipse* rather than a perfect circle. We can draw an ellipse by sticking two pins in a piece of cardboard, placing a loop of string around the pins and the point of a pencil, then moving the pencil to draw the ellipse while keeping the string tight. Fig. 2.8 shows how this might look after placing the pins at A and B with the pencil at point C. Based on the definition of an ellipse, the sum of the distances AC and BC remains constant. Also, as points A and B move closer together, the ellipse becomes more and more like a circle. Now, imagine the moon traveling in an elliptic orbit around the earth as drawn by the moving pencil (point C). We can have the earth's position at either point A or point B. Although the 'eccentricity' of the ellipse is exaggerated, either point indicates that the moon's distance from the earth should vary - between a maximum of about 252,710 miles to a minimum of about 221,462 miles – in the course of its orbit around the earth.

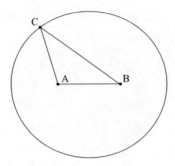

Fig. 2.8. Definition sketch of an ellipse.

Johannes Kepler, a German astronomer, first observed and explained this remarkable fact of planetary motion early in the 17th century. During an **elliptic month** of 27.55 days, the moon passes from a point closest to the earth (**lunar perigee**) to a point farthest from the earth (**lunar apogee**) and back again. Newton's gravity equation predicts that tractive forces will be stronger than usual when the moon is closest to the earth and weaker than usual when farthest away in this orbit. Accordingly, tides will vary from **perigean tides** of greater than normal range to **apogean tides** of less than normal range (apogean-perigean cycle). An increase in tidal range means that high tides will be higher and low tides will be lower. A decrease in range implies the opposite.

2.6 SOLAR TIDES

In addition to lunar tides, Newton's equilibrium theory takes into account the solar tides created by the sun's gravitational attraction and the solar tractive forces acting over the earth's surface as shown in Fig. 2.4 but with the sun in place of the moon. This forcing occurs in concert with the lunar tractive forces, the two acting together through the **principle of superposition.**

At first glance, solar tides could be expected to be larger than lunar tides because the sun's mass is much greater than that of the moon (about 27,100,000 times as great). On the other hand, the sun lies much farther from the earth (about 389 times the earth-moon distance). Looking once again at Newton's gravity equation, we can readily see how changes in the two variables that apply - mass and distance - tend to offset one another. It is then natural to ask: which body has the greater attractive force at a point on the earth's surface, the sun or the moon?

The answer to the above question is - the sun. If we run the numbers through the gravity equation, we find that the sun's attractive force is about 179 times that of the moon. But the question itself is misleading. In our discussion of the tide-producing forces we learned that only tractive forces are capable of producing tides. A tractive force represents only one component of the gravitational force that acts directly toward the moon, or directly toward the sun, and it can be shown that the magnitude of that component varies inversely as the *cube* of the distance from either body. For example, the downward component shown at the 12 o'clock position in Fig. 2.2, previously described as the resultant between centrifugal and lunar attractive forces acting there, would clearly be less for solar attractive forces. Because of the sun's greater distance, the solar attractive force will act along a line more nearly parallel to the earth-sun centerline and therefore offer a closer match to the centrifugal force shown in Fig. 2.2, leaving a smaller residual component acting downward. The end result, while difficult to demonstrate without a lengthy trigonometric development, is surprisingly simple. To compare tractive forces, all we have to do is modify Newton's gravitational equation so that the cube of the distance, R^3, is used in place of R^2. When that is done, solar tractive forces turn out to be less than half (about 0.46 times) those of the moon. Comparing heights, solar tides compare with lunar tides in a similar ratio.

If there were no moon, there would only be one type of tide in the ocean: the solar tide. In a strange world like that our tide observer, Joe, would quickly notice that only 12 hours are needed to complete a solar tidal cycle from one high tide (or one low tide) to the next, or exactly two such cycles per 24-hour *solar day*. But our world does have a moon as well as a considerable number of Joes who have carefully measured a considerable number of tides. Except for a few places with only one high tide and one low tide a day (the Caribbean Sea, for example), the measurements show that two tidal cycles require an average of 24 hours and 50 minutes or exactly one *lunar day*, the same as if there were no sun. This result may seem a little strange since both solar and lunar gravity forces are supposed to be at work at all times. Have solar tides somehow been left out of the daily mix? No, but to witness their effect we have to go beyond daily and look at another semimonthly cycle of the tide: the **spring-neap cycle**.

2.7 THE SPRING-NEAP CYCLE

To illustrate the cycle between *spring* and *neap* tides, another exaggerated view of the earth-moon system similar to Fig. 2.5 is needed, to which we add the sun to make it an earth-moon-sun system (Fig. 2.9). In Fig. 2.9 we're not only taking some license by showing an earth completely covered by water but we further show the moon orbiting the earth in the same plane (the plane of this page) in which the earth orbits the sun. In reality, the earth annually orbits the sun in a plane called the **ecliptic** while the moon's monthly orbit around the earth occurs in another plane inclined at an angle of about 5 degrees to the ecliptic. Overlooking this bit of celestial detail, it is usual to speak of the earth, moon, and sun as being 'in-line' with one another just as the moon reaches the point in its orbit farthest from the sun on the opposite side of the earth (full moon). Then, as shown in Fig. 2.9, lunar and solar tractive forces are also in-line and combine to produce larger tidal bulges (tides of greater range) than either force could produce alone. A repeat of this same situation will happen just as the moon reaches its closest approach to the sun on the other side of the earth (new moon). The tides of greater range that occur during new and full moon are called **spring tides**.

Fig. 2.9. Sun and moon in line with earth producing spring tides.

Tides of lesser range - **neap tides** - occur during the first quarter and third quarter of the lunar cycle when the lunar tractive forces act at right angles to the solar forces as shown in Fig. 2.10. At such times one tidal bulge acts to offset the other.

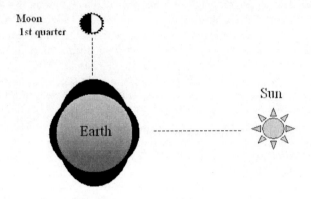

Fig. 2.10. Sun and moon at right angles with earth, producing neap tides.

The moon makes one complete orbit with respect to the sun in 29.53 days, a **synodic month** or the interval required for like phases of the moon to recur. As you can see from Figs. 2.9 and 2.10, two spring tides and two neap tides should occur during this interval. This is indeed what is observed in the form of an increase in tidal range during spring tides, followed by a corresponding decrease during neap tides, thus removing any doubt about the sun's importance in producing tides in the ocean.

Summing up - So far we have discussed three important cycles in addition to the daily rise and fall of water that we call the **astronomical tide**. These are the *tropic-equatorial cycle*, *apogean-perigean cycle* and *spring-neap cycle*. Of the three, the spring-neap cycle is perhaps the best known since it has a recurrence interval of only 14.76 days (it takes half a synodic month to go from one spring tide, or one neap tide, to the next). We are thus more aware of it. The apogean-perigean cycle, however, has a much longer recurrence interval (27.55 days). It too causes a regular variation in tidal range but we don't witness it quite as often and it is somewhat difficult to distinguish from the spring-neap variations that are going on at the same time. However, if you are a fisherman or a boater who regularly checks the tide tables to see how high or how low a given tide is going to be, you probably have noticed that some spring tides seem to produce greater extremes than others. This happens when a perigean tide 'catches up' with a spring tide and augments the higher range with its own large range. What allows the 'catching-up' is the fact that the periods of the monthly cycles involved are slightly different – 27.5 days for the perigean-apogean cycle as compared to 29.5 days for a pair of spring-neap cycles. When the first cycle is finished, the second one still has two days to go; After two 27.5-day cycles, the difference becomes four days and so on.

The third cycle producing monthly tropic-equatorial variations in the tide is the *lunar declinational cycle*. As previously discussed in Sec. 2.5, this cycles produces variations affecting the height of only one high tide or one low tide each day (see Fig. 2.7). The tendency for this to happen reaches a maximum twice during each tropical

Sec. 2.8] **Long term variations in the tide** 19

month of 27.32 days, once when the moon is at maximum declination north of the equator and once when it is at maximum declination south of the equator. Between these times when the moon is on the equator there is little or no inequality in the heights of a day's highs or lows. The recurrence interval for this diurnal (daily) inequality is therefore half a tropical month or 13.66 days. Table 2.1 contains a summary of these important intervals.

Table 2.1. Summary of tidal cycles and recurrence intervals

Cycle	Maximum	Minimum	Interval
spring-neap	range (spring tides)	range (neap tides)	14.76 days
perigean-apogean	range (perigean tides)	range (apogean tides)	27.55 days
lunar declination	ineq. (tropic tides)	ineq. (equatorial tides)	13.66 days

2.8 LONG TERM VARIATIONS IN THE TIDE

The equilibrium tide model is also quite useful when it comes to explaining some important tidal variations of lesser magnitude that last longer than a month. Among these are the *seasonal tides* whose recurrence intervals match certain semiannual and annual variations in solar declination and distance, variations very similar to those already discussed for the moon. However, seasonal tides for all practical purposes would be too small to worry about if it were not for one thing. Depending on the location, they aren't just caused by solar gravity. They get a boost from other effects that follow the sun's seasonal cycles such as solar heating of the oceans and changes in circulation in the atmosphere. As a result, water levels in coastal regions tend to be higher or lower within a slowly varying cycle depending on the time of year. In lower Chesapeake Bay, for example, monthly mean water levels are about 19 cm (7.5 in) higher in September than they are in January of an average year.

When comparing solar declinations and distances with those of the moon, we do have to keep in mind a hard-won point of celestial mechanics (the one that got Galileo in trouble with the authorities in Rome). Unlike the moon, the sun does not travel in an orbit around the earth. Having said that, we can do what mariners have done for centuries and imagine an **apparent sun** that, to an observer on our rotating earth, appears to travel with the moon and stars along the inner surface of a large imaginary sphere centered on the earth. A **celestial sphere** of this type is shown in Fig. 2.11. Note that extending the earth's equatorial plane until it intersects the celestial sphere forms the *celestial equator*.

This model of the heavens is essential for comparing the relative motions of the moon and sun from an observer's point of view. As everyone knows, two of the quarter points of our seasons are marked by the **winter solstice** (sun's maximum declination south of the celestial equator) and the **summer solstice** (sun's maximum declination north of the celestial equator). The other two occur at the **vernal equinox** (apparent sun rising above the celestial equator to the north) and the **autumnal equinox** (apparent sun descending below the celestial equator to the south). Fig. 2.11 illustrates how the apparent sun's motions occur within a plane called the **ecliptic** that is tilted by 23½° with respect to the equatorial plane. The line formed by the intersection of the ecliptic with the equatorial plane has special meaning. One end of this line, the vernal equinox, serves as a reference for the motions of all bodies represented on the celestial sphere.

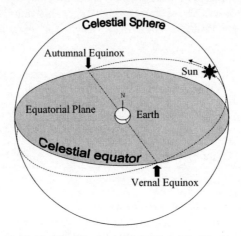

Fig. 2.11. Celestial sphere showing the orbit of the apparent sun.

According to universal custom, the vernal equinox serves as a reference point fixing the orientation of the ecliptic and equatorial planes relative to the stars and their constellations (relative to the *First Point of Aries* in the present age). Like the moon, the apparent sun reaches two maximums in declination during one orbital cycle, a full year in this case. The recurrence interval for the first type of seasonal tide is therefore half a year. Like the moon, the earth's orbit around the sun is an ellipse with a point closest to the earth (*perihelion*) and a point farthest from the earth (*aphelion*). This sets the recurrence interval for the second type of seasonal tide at one year.

Are there tidal cycles longer than a year? Yes and that's another reason why we use the celestial sphere concept. Adding the moon's orbit to the celestial sphere, Fig. 2.12 shows how it appears relative to the plane of the ecliptic:

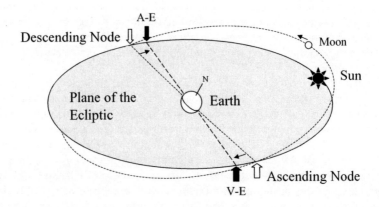

Fig. 2.12. Orbit of the moon relative to the ecliptic and the equinoxes.

The plane of the moon's orbit tilts at an angle of about 5° relative to the plane of the ecliptic (not the equatorial plane). Noting that the equinoxes in Fig. 2.11 serve as reference points fixing the apparent sun's orbit relative to the stars, the lunar orbit is in turn fixed by a pair of points known as the **lunar nodes**. The point at which the moon crosses the ecliptic while rising toward north declination is termed the **ascending node** while the point at which it crosses the ecliptic while falling toward south declination is called the **descending node**. A key fact of celestial mechanics is that the nodes are not fixed relative to the stars. Fig. 2.12 shows them approaching the equinoxes as they slowly move, or regress, toward the west. This westward cycle, known as the regression of the lunar nodes, completes a full 360° in 18.6 years.

Assuming the 18.6-year lunar node cycle starts when the ascending node coincides with the vernal equinox (V-E), you can see that the moon's maximum declination north or south of the equator will be 5° greater than that of the sun during the summer solstice. To show this more clearly, Fig. 2.13 contains two side views of the ecliptic sighting along the vernal equinox.

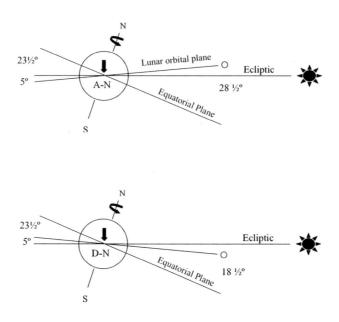

Fig. 2.13. Extremes in lunar declination during the 18.6-year regression of the lunar nodes.

The upper panel in Fig. 2.13 shows the moon reaching a maximum declination of 28½° in its orbit north and south of the equator at the start of the 18.6-year cycle while the lower drawing shows that a minimum north and south declination of 18½° will take place half a cycle (9.3 years) later. These variations modify the daily heights of the high and low tides so that the tidal range varies by a slight amount from one year to the next. This kind of change becomes important when calculating average tidal conditions such as the mean range. For this reason the mean range of all the tides observed at a given place in, say 1990, will differ significantly from the mean range for the tides observed at the same place in 1999.

2.9 CALCULATING THE LUNAR TIDE PRODUCING FORCE

Here is the detailed calculation of the moon's tide producing force that was stated in Sec. 2.2 to be approximately 0.0000034 times the earth's gravitational force acting on a body standing on the earth. The former is understood to act on the body in the direction toward the moon and the latter is understood to act on the body in the direction toward the earth's center. We begin with Newton's law of universal gravitation introduced in Sec. 2.2:

$$F = G\frac{m_1 m_2}{R^2} \qquad (2.1)$$

We now apply this law to an object on the earth's surface with unit mass; i.e., $m_1 = 1$ in equation (2.1) above. In addition, let the following definitions apply:

F_M = force on the object due to moon's gravity

F_E = force on the object due to earth's gravity

m_M = mass of the moon

m_E = mass of the earth

R_M = distance from object to moon's center

R_E = distance from object to earth's center = the radius of the earth

Using these definitions we can now write two equations using the universal law:

$$F_M = G\frac{m_M}{R_M^2} \qquad (2.2)$$

$$F_E = G\frac{m_E}{R_E^2} \qquad (2.3)$$

Dividing equation (2.2) by equation (2.3) and multiplying both sides by F_E we have:

Sec. 2.9] Calculation of the lunar tide producing force

$$F_M = F_E \frac{m_M R_E^2}{m_E R_M^2} \tag{2.4}$$

Introducing Newton's second law of motion, we have

$$F = ma \tag{2.5a}$$

or

$$F_E = g \tag{2.5b}$$

where g is the acceleration due to earth's gravity. Finally, substituting (2.5b) in equation (2.4) yields

$$F_M = g \frac{m_M R_E^2}{m_E R_M^2} \tag{2.6}$$

As previously stated, the ratio of the earth's mass to that of the moon is approximately 81.5 and the ratio of the moon's distance from earth to the radius of the earth is approximately 60.3. Substituting these numbers in equation (2.6) we have

$$F_M = g \left(\frac{1}{81.5}\right)\left(\frac{1}{60.3}\right)^2$$

$$= 0.0000034 g$$

We can save some zeroes by writing the same result using scientific notation:

$$F_M = 3.4 \cdot 10^{-6} g$$

In place of the English system of measurements, scientists generally use the metric system: the meter–kilogram–second or MKS system, for example. The acceleration due to earth's gravity is about 32 ft/sec^2 or $g = 9.8$ m/sec^2 in metric units. The lunar force on a unit mass object would then be

$$F_M = (9.8)(3.4) \cdot 10^{-6}$$

$$= 3.3 \cdot 10^{-5} \, kg.m/sec^2$$

$$= 3.3 \cdot 10^{-5} \, newtons$$

Component forces – Thus far in Sec. 2.9 we have compared the *magnitude* of the lunar tide producing force with the earth's gravitational force on a body without regard to the *direction* in which the forces act. To understand how tide-producing tractive forces arise, we must include direction and examine the component forces through vector analysis. Force vectors indicating magnitude and direction can be represented by their components in a two-dimensional, x-y coordinate system as shown in Fig. 2.14:

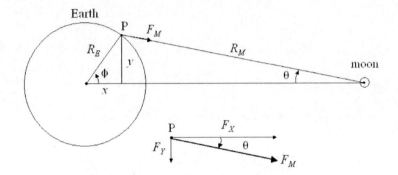

Fig. 2.14. The lunar attractive force, F_M, resolved into components F_X and F_Y.

Using the geometric relationships shown in Fig. 2.14, F_Y can be written as

$$F_Y = F_M \sin \theta \tag{2.7}$$

and since

$$y = R_M \sin \theta = R_E \sin \phi$$

then

$$F_Y = F_M \left(\frac{R_E}{R_M} \right) \sin \phi \tag{2.8}$$

where ϕ is the latitude of the body on which F_Y acts at point P. Substituting equation (2.6) in (2.8) yields

$$F_Y = g \frac{m_M R_E^3}{m_E R_M^3} \sin \phi \tag{2.9}$$

in which we note the cubic exponent has replaced the square of the distances R_M and R_E. We also note that equation (2.9) expresses the differential between the Y force component acting on the body at point P and the Y force component at the center of the earth, which is zero (see Sec. 2.2). Finally, the tractive force due to F_Y is found by projecting F_Y onto a line tangent to the earth at point P. A similar development obtains the F_X component whose projection onto the tangent at P must be added to that of F_Y.

3

Are tides waves?

Yes. Most of the practical information about tides - information that gets put to use when predicting how high the tide will be or how strongly the tidal current will flow at a certain time and place - depends on treating the tide as a special kind of wave. In this chapter we will do just that with one idea in mind. That is, the more you know about waves, the better you will understand tides, which are a special kind of wave.

Before we begin, a word of caution about the familiar words *tidal wave*. This is a perfectly good term but one sometimes used (mistakenly) to describe a **tsunami**, a highly destructive sea wave generated by an undersea earthquake, landslide, or volcanic eruption, not by the tidal forces described in Chapter 2. The misuse of this phrase is somewhat unfortunate. No one is alarmed when told that the tide is entering the bay but if you say 'tidal wave' many will start for the hills thanks to certain popular films.

It may sound like mere quibbling over words but we already have oversimplified terms like 'tidal bulge' that tend to stick in everyone's mind. Getting beyond these almost too familiar concepts has its rewards. As shown in Chapter 2, the equilibrium theory coupled with the tidal bulge concept is very useful in understanding how certain types of tide occur but it has a problem. Parts of the concept are literally untrue. The earth is not completely covered by water and there is no such thing as a pair of tidal bulges traveling completely around the earth. Equally important, real tides do not respond instantly to the tide-producing forces of the moon and sun, as the equilibrium theory requires.

3.1 AN ENCOUNTER WITH WIND WAVES

Even on a day when the ocean's surface appears 'flat calm', it would be unusual if no waves at all were present, excluding the tide. If we happen to be on a ship far from land, the tip-off as to the presence of waves might come as our vessel gently rocks from side to side although there's no visual clue that anything is going on in the water around us. Probably we are experiencing *swell* - waves that have reached our position after traveling a great distance from some other part of the ocean we're in. We hardly notice the slight rise and fall of the glass-like water surface because there's no vertical reference, like a piling stuck in the bottom nearby, and the rhythm of the motion is slow. Using a stop watch to determine the number of seconds it takes for the vessel to rock back and forth, say fifty times, we could divide the number of seconds we get by fifty and obtain an estimate of the average wave period. The answer would probably lie somewhere between 10 and 18 seconds. This is easy to take so let's stay a little longer on board our ocean vessel.

The sky has begun to change and a light wind has come up. The wind's stress on the water creates small ripples so that the surface no longer appears glassy. As the wind speed increases to about 8 knots[1], the ripples combine to form wavelets that grow in height until their tops begin to break off creating whitecaps here and there. At 22 to 27 knots the wavelets merge, grow and become full-sized waves. There are whitecaps everywhere and spray from breaking waves enters the air. While this sea state can form pretty quickly (perhaps an hour or less depending on how rapidly the wind speed increases) it does not happen instantaneously. A certain amount of time is needed to transfer energy from the wind to the waves. The waves 'grow' as a result of a complex interaction between wind pressure and wave shapes emerging out of the continually changing water surface area. Waves coupled to the wind in this way are termed **wind waves** but they belong to a larger class of waves called **forced waves**. The characteristics of these waves remain constant only as long as the forcing mechanism, the wind in this case, remains constant. But suppose instead the wind has increased and the waves have grown larger in our make-believe journey.

3.2 A SPECTRUM OF WAVES SUPERPOSED ON THE SEA

We've been in a storm and some pretty large waves have rocked our vessel much more vigorously than anything we've experienced up until now. The stopwatch has long since been put away. The wind, after reaching a peak and remaining strong for many hours, has created a fully developed sea: in short, a real mess. Describing its characteristics in terms of the individual waves present is seemingly impossible. In fact it would be without the **principle of superposition**.

The superposition principle tells us that, no matter how complex the surface of the water appears to be at any given instant (ignoring the spray and the whitecaps), it can be broken apart into a collection of simple sine waves (Fig. 3.1), each one distinguished by a **height** (vertical distance from trough to crest), a **length** (horizontal distance from crest to crest), and a **phase** that marks the position of the waveform in horizontal distance and time.

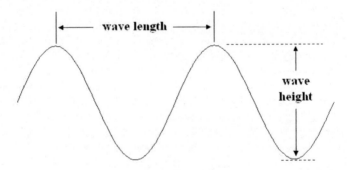

Fig. 3.1. Sine wave illustrating wave height and wavelength.

Moving waves have a fourth attribute, the **wave period** or time required for the recurrence of like phases at a fixed point and a fifth attribute, *wave direction*, arises in three-dimensional space. Starting with a completely calm sea, we could in theory add component waves to it one at a time, aligning the mid-level between wave crest and

---

[1] 1 knot = 1 nautical mile per hour = 0.51 m/sec

wave trough of each wave with the still water level. Superpositioning means that at every point on a horizontal grid, the final elevation of the water's surface (measured above or below still water level) must at any given moment equal the sum of the individual heights from the wave *ensemble* at that point.

The term 'random sea' implies a wave ensemble displaying a wide range of values among the five attributes identified above. Spectral analysis provides a means of estimating the distribution of those values for a given place and time but it's a big ocean and the necessary wave records are available in relatively few places. And if we're everyday mariners aboard a ship overlooking the jumbled water masses in motion, we won't have that option. We may see an occasional high point as the water rises up, calling that a crest, and about as often notice a low point as the water falls away and call that a trough. If (a large if) we have good sense of scale, we might estimate the vertical distance between a number of the crests and troughs we see and make an estimate of the average wave height. Ship's officers and crew for many years have done just that (often using a part of the ship for scale) so that naval and maritime organizations could compile the data and provide sector maps describing the *wave climatology* in oceans and seas around the world. Studies have shown that shipboard observers tend to count the highest waves (the lowest waves are more numerous and harder to distinguish), causing their 'average' wave height estimates to be biased toward the high side. This is not an unwelcome result to those for whom the higher waves are the ones of interest, not to mention the offsetting argument that smart captains tend to avoid areas with large waves.

Ocean engineers employ a similar bias when using instruments that automatically record the height of every wave encountered at a fixed location over an interval of several minutes. They rank the recorded waves in order according to height, then average only the highest one-third of the waves and call the result the **significant wave height**. Coupled with the detailed knowledge that ocean scientists and engineers have learned about the distribution of individual wave heights in the sea (described by the Rayleigh distribution in the field of statistics), the significant wave height has proven to be a very useful measure for a number of engineering tasks; for example, the estimation of maximum wave height for a given sea state based on a sampling of the waves.

Other visual characteristics of ocean waves may emerge after examining them a little more closely. Wave crests tend to have more noticeable peaks and troughs tend to be flatter. Sometimes (as when looking down on a single wave train from an aircraft) we can make out a rough pattern of lines connecting some of the crests. The lines are neither long nor straight and we speak of these wind-generated waves as being *short-crested* and *irregular*. Still, the eye can usually pick out a general direction toward which the wave crests appear to be moving and specify the average wave direction: usually the direction in which the wind has been blowing unless the latter has changed considerably with time. The wave direction itself is by no means uniform; instead there is a fan-like spread on either side of a mean direction describing the general advance of the waves we see. As the wind continues to die away, so do the highest waves around our ship and the character of the sea begins to change again.

Without wind forcing, the wind waves described above become **free waves** and begin to leave the area where they were generated. A remarkable thing happens at this point. We didn't notice it of course during the storm but the jumble of waves that were created varied not only in height but in period as well. Now imagine cutting a vertical slice through the sea surface in the direction of wave advance. The resulting wave profiles –

crests and troughs and the surface in between – would appear to travel away from us at a certain speed. Engineers refer to this speed as the **wave celerity** or **phase speed** and use the letter C_o as its symbol in deepwater areas. A simple equation then applies:

$$C_o = \frac{g}{2\pi} T \qquad (3.1)$$

where g is the acceleration of earth's gravity, T is the wave period, and π is a universal constant[2]. Celerity, by the way, is not another word for current. The term *current* applies to the average speed of water particles flowing in a specified direction. Celerity refers only to the speed of an individual wave profile propagating at the water's surface; the water itself is not moving at that speed. A 6-second wave, for example, has celerity of about 9 m/sec (31 ft/sec). A 12-second wave, on the other hand, travels twice as fast and a 24-second wave moves away from the generating area at four times this rate. Waves born in the same crib, so to speak, quickly disperse from one another because some travel much faster than others. Somewhere far away another vessel like ours may soon be sitting on a glassy sea experiencing our swell.

Free waves traveling across the ocean, particularly the ones with shorter periods tend to decay with time because of frictional dissipation and will eventually disappear. Those that encounter land in the meantime will have a different fate. As the depth becomes shallower approaching land, **shoaling** and **refraction** transform the waves in terms of their height and direction while leaving the wave period basically unchanged (more on this shortly). *Long-crested waves* result, the kind that every surfer hopes to encounter at the beach. It is now a little easier to observe the wavelength directly as the distance between successive crests. Wavelengths are harder to distinguish in deep water but it is easy to estimate them if we know the wave period and thus the wave celerity, C_o. This is because a moving wave must travel one wavelength during one wave period so that

$$C_o = \frac{L_o}{T} \qquad (3.2)$$

Substituting equation (3.2) in equation (3.1) and multiplying both sides of the resulting equation by T, we obtain the formula for the deepwater wavelength as

$$L_o = \frac{g}{2\pi} T^2 \qquad (3.3)$$

A question may be on your mind at this point. We've been talking about waves in 'deep water'. When does this term cease to apply; i.e., how do we know when a wave is in 'shallow water' and what happens then to descriptive characteristics like wave height and wave length? The handiest measure describing deepwater waves is the wavelength, which we get from equation (3.3) after timing the waves to find their period. Water

---

[2] Ratio of a circle's circumference to its diameter; $\pi = 3.1416$ approximately

depth from surface to bottom can be measured over an entire ocean but there's no obvious cutoff point between deep and shallow water. If I were thinking about swimming, I'd say it's deep water when it's over my head. Instead, we are talking about waves in the ocean so why not compare the depth to the length of the wave? Passing over the physics involved, it turns out that in order for the above formulas for C_o and L_o to be correct, the depth must be at least one half of the wavelength for a **deepwater wave**. A quick calculation[3] for a 6-second wave shows that $L_o = 56.2$ meters (184 feet) is its deepwater wavelength and we note that it applies only in water depths equal to at least one-half of this value (i.e., about 28 meters or 92 feet).

As waves leave deep water and travel inshore toward shallower depths they undergo a transition. Celerity and wavelength no longer depend on the wave period alone but depend more and more on the water depth as the latter continues to decrease. When we finally do arrive in 'shallow' water, wave celerity no longer depends on wave period at all but depends entirely on the depth as stated in the following equation:

$$C_S = \sqrt{gh} \qquad (3.4)$$

where h is the water depth and g is the acceleration of gravity. Since equation (3.2) retains the same form for shallow water, we get the corresponding wavelength for a **shallow water wave** by multiplying its celerity by the wave period:

$$L_S = T\sqrt{gh} \qquad (3.5)$$

Again omitting the physics, the above equation is accurate whenever the depth is about one-twentieth of the wavelength or less. This relationship explains why waves 'slow down' and their wavelengths get shorter and shorter as they approach the beach. It also explains why waves undergo *refraction* in shallow water; If the crest line of an individual wave approaches the beach at an angle to the depth contours, the segments of the wave in shallower water inshore will slow down relative to the offshore segments in deeper water, causing the crest line to refract or bend so that it tends to become parallel to the underlying depth contours. As refraction takes place, changes in wave direction can be visualized by imagining the change in a series of evenly spaced lines drawn normal to the advancing crest line.

Shallow water waves moving inshore also become steeper (their height-to-length ratio increases) as wavelength shortens and wave height increases due to *shoaling*, a consequence of wave power conservation. Shoaling waves steepen until they become unstable and break (some waves partially reflect at the shore and go back to sea). Energy transported by waves that may have traveled hundreds of miles is suddenly released within a narrow zone next to the beach.

3.3 A LOOK INSIDE THE WAVES

So far we have only discussed the characteristics of waves as we see them at the surface. It's time to draw a picture of a wave and talk about what lies beneath the surface: the anatomy of an individual wave. The ideal configuration for the wave

---

[3] Use $L_o = 1.56\ T^2$ (meters) or $L_o = 5.12\ T^2$ (feet) where T is in seconds.

profile is provided by the sine wave from elementary trigonometry. A definition sketch of a sine wave is shown in Fig. 3.1.

Let's bring back "Joe the Observer" and place him on a ship far out in the ocean in deep water. Some medium-sized swell is passing by and Joe has a sea anchor deployed so that his ship tends to remain in one position with respect to the bottom far below. As the waves pass his ship, Joe sights along their crest line and happens to see a sea bird resting on the water. Note the word 'resting' – this bird is totally inert and fully coupled to the water. Joe observes a few more waves, noting how the bird moves up and down as each one passes beneath it. On closer inspection, it's apparent that, besides moving up and down, the bird also seems to move forward with each wave when over its crest and to move backward (away from the direction of wave advance) when over a trough. Pretty soon he is satisfied that the bird's motion describes a circle whose diameter equals the wave height. Fig. 3.2 depicts what Joe would see during the passing of one of these waves, marking the sea bird's position with a small dot on the circle.

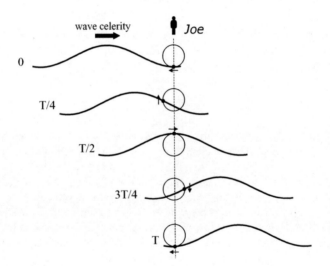

Fig. 3.2. Snapshots in time of a single wave passing Joe the Observer.

The above figure shows five separate positions of a single wave as it moves from left to right, the corresponding times for each relative position (wave phase) differing by a quarter of the wave period. Note that as the fifth position is reached at time T, the wave form has returned to the same phase it had at time zero. After one wave period, both the sea bird and the water particles directly beneath it have traveled through one complete circle (a clockwise orbit looking at waves moving to the right) and are back at the point in the wave trough where they started the cycle [4]. This much is apparent to anyone following a point on the water's surface and allowing for some irregularity.

---

[4] Or a little beyond this point. A slight amount of net water movement 'down-wave' with each orbit is known as *Stokes drift*.

What about particles below the surface? Clearly particles just below the surface can't remain still while those at the surface are moving up and down, forwards and backwards, in line with the wave's direction of advance. An experimentalist might try placing tracers in the water, say a few ping-pong balls filled with a liquid of similar density to make them neutrally buoyant. After persuading Joe to become a diver and extend his observations underwater, he would report that these objects, allowing for some irregularity due to turbulence, also move in a circular orbit induced by the waves. And after extending his dive downward while in deep water, he would add that the diameter of the orbits decreases rapidly with depth as shown in Fig. 3.3. Orbital diameter in fact decreases exponentially with depth in deep water, reaching about 4 percent of its surface value at a depth equal to one-half the deepwater wavelength.

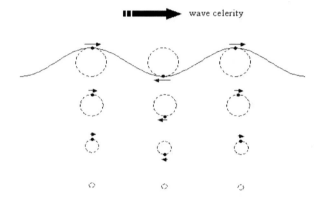

Fig. 3.3. Orbital motion of water particles inside a deepwater wave ($h/L=1/2$).

Once the deepwater wave shown in Fig. 3.3 moves into a shallower region, its wave depth-to-length ratio becomes less than 0.5 and it is then classified as an *intermediate wave* with characteristics in between those of deepwater and shallow water waves. One of the results of this transition is that the circular orbits characteristic of deepwater waves begin to degenerate into elliptical orbits as the vertical axis of the orbit decreases and the horizontal axis increases. The wave becomes a shallow water wave (Fig. 3.4) as it moves into depths less than one twentieth of the shallow water wavelength, L_s, as calculated by equation (3.5).

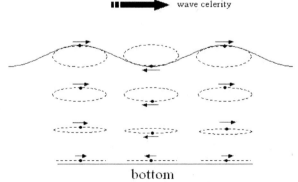

Fig. 3.4. Orbital motion of water particles inside a shallow water wave ($h/L=1/20$).

As illustrated in Fig. 3.4, the length of the horizontal axis of each ellipse within a shallow water wave remains constant from top to bottom. However, the length of the vertical axis gradually shortens from a maximum at the surface to zero length at the bottom where water particle motion is restricted to horizontal movement back and forth. This motion due to the wave causes a stress on the sea bottom much as the wind initially transmitted a stress on the sea surface to form the wave. Instead of more waves, however, bed erosion is the most likely result if the wave stress exceeds a critical limit.

3.4 TIDES AS LONG WAVES

It's now time to discuss tides as water waves which can be measured using a special type of wave gauge called a **tide gauge** (*tide gage*). Tide gauges are instruments that measure and record tidal heights at regular intervals of time at a fixed location. The records they produce contain waveforms very similar to the sine wave shown in Fig. 3.1. The difference is that the wave periods for tides are measured in hours, not seconds. For that reason, tide gauges generally require some mechanism, such as a **stilling well**, for filtering out wind waves that would otherwise 'contaminate' the tidal records.

Tidal heights and wave heights are similar enough in range to use comparable vertical measurement scales in meters or feet. But, as we learned in Chapter 2, the period of the semidiurnal lunar tide is 12.42 hours. At an ocean depth of 5,000 meters, its shallow water wavelength, L_s, would be about 9,900 kilometers or nearly a quarter of the distance around the earth. In 5 meters of water, L_s would be about 313 km or approximately the axial length of the Chesapeake Bay. These results clearly put the tide in the category of an ultra-shallow wave or **long wave** anywhere on earth. Fig. 3.5 illustrates a long wave with its water particle ellipses stretched to the point where there is little vertical motion anywhere as compared to the long, horizontal stroke of the water particles oscillating inside the wave (considerable vertical exaggeration is needed just to show the crests and troughs in this figure). This horizontal motion is the **tidal current** that is termed **flood current** when moving in the direction of wave advance and **ebb current** when moving in the opposite direction. Notice that the currents reach maximum speed under the crests and troughs of these idealized waves and reach zero speed or **slack water** at points halfway in between - points where the orbital motion would be either up or down inside a deepwater wave.

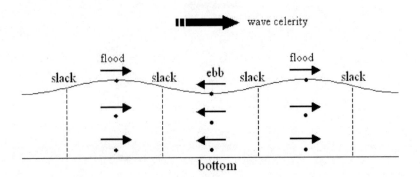

Fig. 3.5. Horizontal motion of water particles inside a long wave in shallow water.

3.5 OCEAN TIDES AND KELVIN WAVES

Apart from having very different periods and wavelengths, a key characteristic setting tide waves apart from wind waves is the fact that the tide is always a forced wave and never a free wave. Although winds can produce a long wave oscillation called a **seiche** in large lakes and certain other bodies of water, a different type of forcing is responsible for tides in the ocean. You already know about this force from the discussion of lunar and solar tractive forces in Chapter 2. The interesting part, the part we are going to look at now, is how these forces are applied to generate tide waves in the real ocean – the six oceans and approximately seventeen seas scattered over the earth. Each of these water bodies is shaped more or less like a circular (or rectangular) basin rimmed by land or part of an adjoining ocean or sea. Not all are connected and there is no world-circling ocean near the equator where a tide wave (tidal bulge) could follow the moon and apparent sun traveling east to west without interruption[5]. An unusual wave is needed that can operate within the confines of these large basins and still produce the tidal periods and other features characteristic of the tides we observe on a day-to-day, month-to-month basis. The kind of wave that fills the bill is called a **Kelvin wave** or for those who prefer less technical names, a **rotary wave**.

What is a Kelvin wave and how do we create one? In its simplest form, it's a wave that travels within a fixed region and oscillates in two horizontal directions rather than one. Strictly speaking, only large-scale forces can produce a Kelvin wave. But you create a mock Kelvin wave every time you swirl your coffee or tea in a cup. If it's a clear glass cup, you can look through the side and see a crest, then a trough, and then a crest sweep by again as in Fig. 3.6. You, of course, create this wave by means of a circular motion with your hand, a regular *periodic* motion you impart without even thinking about it. Now – keeping the same process in mind – transfer this image from your cup to the North Atlantic ocean. As the crest of the wave passes the coast of Virginia, we would call that a high tide. A little more than six hours later that same high tide might swirl past the coast of England and continue on to reach Virginia again after a circuit lasting twelve hours and 50 minutes all together. Be careful not to swirl too hard or you'll flood Virginia!

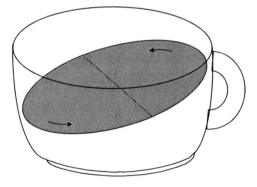

Fig. 3.6. A mock Kelvin wave in a cup of tea.

---

[5] A shallow-water wave would required a depth of 22 km to keep up with the moon.

Given the above cup-in-hand analogy, we are pretty close to Galileo's idea of a barrel of water on the deck of a moving ship mentioned in Chapter 2. But again, it is the lunar and solar gravitational forces, not the earth's motion, that are responsible for creating tides. Just as the wind's stress on the sea surface created the wind waves discussed earlier in this chapter, the invisible net of tractive forces illustrated in Fig. 2.4 sweeps over the spinning earth to create the long waves we call tides. But, unlike wind, which imparts a stress only on the water's surface, the lunar and solar tractive forces act on every particle of water in the ocean. So although the tractive forces are very small, they act on huge masses of water, masses that are free to deform and take the shape dictated by the tide. Also unlike the wind, these forces never cease and keep the same rhythm indefinitely, which is the reason for calling the tide a forced wave.

All of the variations in the tide that were explained in Chapter 2 using the 'tidal bulge' concept can be explained as well by the net of tractive forces acting over a spinning earth. As the earth turns on its axis, individual oceans and seas will feel the pull of the tractive forces, first in one direction and then the other. We will analyze a small part of that motion involving waves that progress and those that appear to stand still to learn how the Kelvin wave is set up. First we'll look at a different kind of wave called a **standing wave** and examine its behavior.

3.6 PROGRESSIVE WAVES AND STANDING WAVES

Waves that have a forward speed (wave celerity) are called **progressive waves**. When a train of these waves hits a reflecting boundary such as a sea wall, they completely reverse their direction of travel and immediately combine with incoming waves that have not yet reached the boundary. The combination of the two wave trains traveling in opposite directions results in a standing wave that, to an observer on shore, does not appear to move in any direction. This raises an important question - exactly how do the wave trains 'combine'? According to the ordinary principle of superpositioning, one wave form is simply placed on top of the other at a given instant and matching surface elevations are added together, producing a single wave in their place. The elevations that are added are not the elevations relative to the bottom but the surface elevations relative to the still water level when no waves are present, a level usually approximated by the **mean water level** (*MWL*) in a water level record. Reckoned this way, surface elevations are in turn both positive and negative.

Consider an incoming wave and a reflected wave with the same height and length in a rectangular basin open at one end (Fig. 3.7). When *in phase* (crests matched with crests, troughs matched with troughs) the two waveforms will combine to produce a single wave with twice the height of the contributing waves. When *out of phase* (crests matched with troughs) the two waves offset one another resulting in a level surface coinciding with the mean water level.

We can also generate a standing wave in a cup of tea by moving it back and forth instead of swirling it around. The water's surface would look exactly like the tilted surface shown in Fig. 3.6 except that, rather than rotating, it will oscillate up and down on either side of the dashed line in the figure. This line corresponds to the **nodal point** N shown in Fig. 3.7. No vertical motion occurs along the line so long as the cup is moved back and forth without swirling. Looking again at Fig. 3.7 we can see how a standing wave like that in the cup might be created by simply placing a second barrier one-half of a wavelength away from the first as shown in Fig. 3.8. The basin then has no incoming waves but is nevertheless quite open to forcing inside its boundaries.

Sec. 3.6] **Progressive waves and standing waves** 35

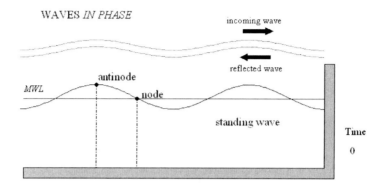

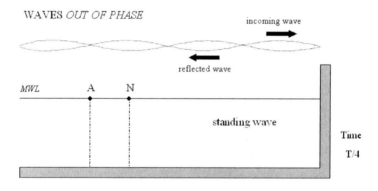

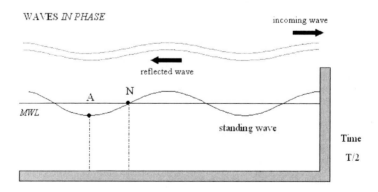

Fig. 3.7. Standing wave created by superpositioning of incoming and reflected progressive waves in a basin open to the sea.

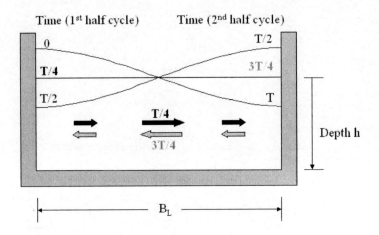

Fig. 3.8. Standing wave in a closed rectangular basin.

Fig. 3.8 shows standing wave positions occurring during a single wave cycle of period T in a closed hypothetical basin of length B_L. The fundamental oscillation in this type of basin occurs when the basin contains half a shallow water wavelength or

$$L_S = 2B_L \tag{3.6}$$

although higher 'modes' are possible with

$$L_S = \frac{1}{n} B_L \; ; n = 1,2,3,...$$

Substituting the wavelength from equation (3.5) into (3.6) yields

$$B_L = \tfrac{1}{2} T \sqrt{gh}$$

which gives the free period of oscillation in the basin as

$$T = \frac{2B_L}{\sqrt{gh}} \tag{3.7}$$

The period given by equation (3.7) is the basin's **natural period** at which the water inside will oscillate, if periodic forcing is somehow provided that matches or comes close to that period. Your hand movement matches the natural period for your teacup more or less by trial and error. For tides oscillating in a large basin, the natural period of the basin must be somewhere near the dominant period of the astronomical tide (12.42 hours or, in some cases, 24.84 hours).

3.7 WAVE DYNAMICS AND WAVE ENERGY

What governs the height of a standing wave? The height of the wave and the way it is produced leads to the topic of wave dynamics. Wave motion in a tidal basin can be compared to a simple pendulum swinging back and forth. Both have a characteristic length that determines the period of oscillation, which in the case of a pendulum of length L is

$$T = 2\pi \sqrt{\frac{L}{g}}$$

and both have an **amplitude** or maximum distance through which motion occurs to either side of the 'neutral' position (mean water level, vertical pendulum). Amplitude when expressed this way is equal to one-half of the standing wave height in the basin. Let's look at the pendulum first to get a clear idea of the dynamics involved.

Suppose you are pushing a person in a swing made with a pair of ropes attached to a tree limb. If the limb is, say, 9 meters (30 feet) above the swing seat, the natural period will be about 6 seconds. This means that if you stand in one place and give the swing a push, you have to wait 6 seconds before you push again if you want to keep the swing in motion at its natural period. Meanwhile, if the person wants to go 'higher' you must push harder (provide a bigger impulse) to increase the amplitude of the swing. Why? The added impulse is needed to overcome the restoring force (the component of gravity force, or weight, acting toward the neutral position). The maximum restoring force increases as the amplitude increases with the person reaching greater heights above the ground at both ends of the swing's arc.

Of course you want to be careful, not only because the occupant might fall out of the swing but also because there is a limit to how big the arc can be with the swing still behaving as a pendulum. But if you happen to be pushing a heavy adult instead of a child, there's less to worry about; you're going to need a lot more impulse to get much height on a 9-meter swing. Meanwhile the amplitude of the oscillation, as long as the swing is still behaving as a pendulum, has no effect on the period, which remains constant. And for waves in tidal basins, just as for swinging pendulums, the amplitude of each oscillation slowly decays once the forcing is removed. This is due to the effects of friction that are always present in a system with moving parts.

You may have noticed the two sets of arrows in Fig. 3.8. These represent the oscillating currents that are associated with standing waves. No water can flow into or out of the solid barriers at either end of the basin so the current there is always zero. However, rising and falling water levels at either end can only occur if there is a transfer of water out of the half that is falling into the half that is rising on either side of the nodal point. The result is a current that is strongest just beneath the nodal point, diminishing to zero at the antinodes at either end of the basin. In complete opposition to a progressive wave, a standing wave has zero current, or slack water, when all water levels reach their maximum or minimum heights (times 0 and $T/2$ in Fig. 3.8). Currents reach their peak strength when all water levels reach mean water level (times $T/4$ and $3T/4$ in Fig. 3.8).

The 'moving parts' in the systems described above have a very basic requirement. Energy, the capacity for doing work, is needed to start and maintain the motion. In the swing example, the motion is started by either a push or a pull but when the swing reaches the top of its arc, the motion briefly ceases. At that moment, when the swing (its mass and the mass of the person in it) has been lifted as high as it will go, all of the energy in the system becomes **potential energy** waiting to be released. As the motion reverses and the swing reaches the bottom of its arc, all of the potential energy is transformed into **kinetic energy,** energy gauged not by height and the work done against gravity to get it there but by the motion of the swing traveling at maximum speed. At all other times there is both potential and kinetic energy in the system, the sum of which remains constant until more is added (by pushing) or removed (by friction).

The standing wave system shown in Fig. 3.8 also involves a combination of potential and kinetic energy. When the maximum and minimum water levels are reached, there are no currents in the basin and the system contains only potential energy gauged by the position of the water masses above or below the reference level (i.e., *MWL*). When the water levels are everywhere the same, the currents reach their maximum rate of flow and the system contains only kinetic energy. At other times it contains both. The system will lose energy constantly due to internal friction within the water mass and boundary friction along the sides and bottom of the basin. Clearly if there are no permanent outside energy sources, such as the tidal tractive forces oscillating at just the right frequency, the standing wave motion in the basin will gradually die away unless wind or another type of temporary forcing contributes an impulse at or near the natural period.

3.8 KELVIN WAVES IN OCEAN BASINS

Earlier we noted that a semidiurnal tide wave with 12.42-hour period traveling in an ocean 5,000 meters deep would have a wavelength of around a quarter of the earth's circumference. If we could order a tidal basin of that depth to be constructed and filled with ocean water in the hope of getting a good tidal response, what length (or diameter) would we give it? About one-eighth of the circumference or 4,950 kilometers (2670 nautical miles) is obviously a good answer to that question. We might even want to cancel the order because an existing ocean basin with roughly the same dimensions won't be hard to find. The North Atlantic deep ocean basin, for example, is just slightly smaller in both diameter and average depth.

Taking the North Atlantic basin as the example, we will make an initial assumption that it has a standing wave like the one shown in Fig. 3.8 while recognizing that the vertical dimension in this figure is greatly exaggerated. The basin is located in the northern hemisphere starting just above the equator where it will certainly experience the necessary push-pull of the lunar tractive forces as they appear in their globe-circling net in Fig. 2.4.

Examining Fig. 2.4 closely, it can be seen that during equatorial tides when the moon lies in the equatorial plane, the southern part of the basin will feel a simple east-west oscillation in the lunar tractive forces as the earth turns on its axis. Recalling that one turn with respect to the moon defines a lunar day, it is clear that the lunar tractive forces experienced at a point on the equator will vary like a sine wave, reaching two maximums (eastward force) and two minimums (westward force) every 24.84 hours. This is the exact forcing needed to drive the lunar standing wave. The solar standing wave will have similar but weaker forcing over a 24-hour period. Moving north of the equator into the more central regions of the North Atlantic basin, the tractive forces progressively gain a north-south component in addition to the east-west forcing and it is at this point that vector analysis becomes necessary in order to see the details of the breakdown between east-west and north-south tractive force components – components somewhat analogous to winds that continually change their magnitude and direction in a regular but rather complex way. The change in the magnitude and direction of the tractive forces, however, is not at all random but consists of a series of periodic variations that are completely predictable, even when the moon, or sun, is not over the equator but at some other declination north or south.

Now that we understand the forcing a little better than before, it is easier to see that systems like the North Atlantic basin are unlikely to acquire an ordinary standing wave oscillating only in the east-west direction as shown in the upper part of Fig. 3.9. North-south tractive force components suggest a wave oscillating in other directions as well. There is another reason to expect such a result.

The Coriolis effect – Any large-scale horizontal current stream in the northern hemisphere experiences an accelerating force to the right (to the left in the southern hemisphere) because of the **Coriolis effect**, a consequence of the earth's rotation. When we track large water masses in motion, we employ a 'flat map' or reference plane tangent to the earth's surface at our location. That plane moves with the earth and thus acquires a component of the earth's rotation at all latitudes excepting the equator; similarly it acquires a component of eastward linear velocity (relative to the stars) everywhere on earth except at the poles. Representing earth's rotation as a vector, Ω, parallel to its rotation axis, we obtain the local component normal to our reference plane by projection based on the sine of the latitude as illustrated in Fig. 3.9. Linear velocity, on the other hand, varies as the cosine of latitude.

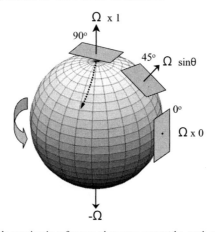

Fig. 3.9. Component of earth's rotation in reference planes tangent to the earth at different latitudes.

It is perhaps easiest to visualize the Coriolis effect at the geographic poles (90°N, 90°S) where the reference plane rotates at the rate of 360° per day but has no linear velocity. In a commonly used example, we may imagine a missile launched at the north pole and note its trajectory to the south. Excluding a downward acceleration due to gravity that keeps the missile in orbit around the earth, Newton's first law of motion ("*Every body continues in a state of rest or of uniform motion in a straight line, unless it is compelled to change that state by forces impressed on it*") requires it to travel in a great circle that does not turn with the earth since it has no eastward linear velocity. Plotting its position on our reference plane, which continues to rotate counter-clockwise to the east, the missile appears to deflect its motion to the right (clockwise west) as if it were responding to a westward force impressed on it. Although the force is fictitious, the effect is quite real. If launched in any direction at latitude 45°N, the missile would again travel in a great circle orbit and would again show a deflection to the right in the horizontal reference plane. However, there would be no deflection for a missile launched at the equator because the reference plane positioned there has no component of rotation parallel to earth's axis and is moving to the east with the same linear velocity as the missile. Clearly we can only represent small regions of the spherical earth on a single reference plane so the above only applies very near the equator.

Given a standing wave caused by tidal forcing in an enclosed ocean basin (Fig. 3.8 and upper panel, Fig. 3.10), Coriolis acceleration causes the water surface to rise to the right of the current direction while falling on the left, creating a new oscillation in the transverse direction. While this process continues through the wave cycle, the currents continually change their direction but not their magnitude, resulting in a rotary wave as shown in the lower panel of Fig. 3.10. Due to the Coriolis effect, the rotation is always in the counter-clockwise direction in the northern hemisphere, clockwise in the southern hemisphere with no effect along the equator.

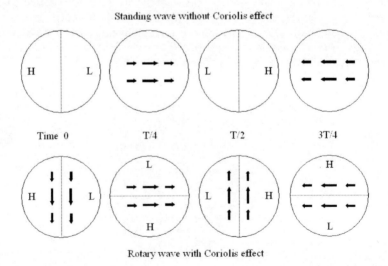

Fig. 3.10. A standing wave compared to a rotary wave in a circular basin.

The rotary wave (Kelvin wave) reveals some interesting properties when compared to a standing wave. A standing wave, for example, produces no change in water level along the nodal line (dashed line, upper panel Fig. 3.10). Although a nodal line is also shown for the rotary wave (dashed line, lower panel Fig. 3.10), this line continually changes its direction by rotating about a single point called the **amphidromic point**, the only point in the basin where water level does not change and there is no tide. Also because of rotation, the water level in a rotary wave does not 'stand' at either high tide or low tide throughout the basin every T/2 hours as it does in a standing wave system. This fact might seem odd when you look at the rotary wave in Fig. 3.10, which, at any given moment is represented by a tilted surface just as the standing wave is. The difference is that, for the rotary wave, the surface is continually rotating about the amphidromic point in the basin center.

To underscore this point, consider that a tide gauge stationed at any point along the rim of the basin will record a sine wave as the tilted plane representing water level continues its rotation (see Sec. 3.9). At times of high or low water at this station, tides at all other locations around the rim will still be rising or falling toward their extremes. In all respects this sine wave behaves as a progressive wave that has been *trapped* by the basin shoreline and forced to make its way around the rim. Islands between the rim and the amphidromic point will record a similar sine wave with a lesser amplitude (lesser tidal range).

This is information that allows a completely new framework for depicting tides through time and throughout a region of the earth's water-covered surface. For example, by looking at the tide as a rotary wave, we now see that it has the same relationship to the basin it operates in as the standing wave shown in Fig. 3.10. That means the tide also shares the same basin size requirement; i.e., the rotary wave must have a basin with the right combination of length (diameter) and depth to produce a natural period near that of the tide involved. If the dimensions come close enough to those required for a 12.42-hour lunar period, then 12.42-hour lunar tides result. Being close to a 12-hour natural period, 12-hour solar tides usually result as well. The North Atlantic ocean basin meets this requirement in general and consequently such tides originate there. Shallow, elongated systems such as the Chesapeake Bay meet the periodic requirement (Equation 3.7) but lack the required mass for significant tidal forcing. As will be shown in Chapter 6, the tide wave created in the North Atlantic ocean propagates into Chesapeake Bay and its tidal tributaries as a long wave. In other words, assuming a wall was built across the Chesapeake Bay entrance, the bay would become almost tideless.

Given what we have just learned, it seems possible to represent the tide in an ocean basin more realistically in terms of the familiar properties of tidal range and phase. We can map all the points that experience the same tidal range using **co-range lines** and all the points that experience the same tidal phase (i.e., high water) at a given time using **co-tidal lines**. Plotting the two together in an ideal basin in the northern hemisphere gives a result such as that shown in Fig. 3.11 below. In the figure, dashed circles represent co-range lines and the radial lines are co-tidal lines at the lunar hour indicated running counter-clockwise around the basin. This of course is an ideal basin, which we could not expect to find with the very same appearance in the real world. The co-tidal lines in particular would be likely to vary both in shape and position because of variations in the shape of the basin and its bathymetry.

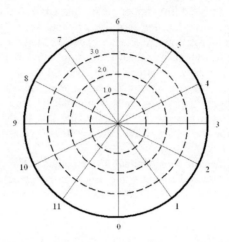

Fig. 3.11. Hypothetical circular ocean basin showing co-tidal and co-range lines.

Unlike tea cups, real ocean basins contain quite a few irregularities. These show up not only in the shape and configuration of a broken and irregular shoreline but in the distribution of depths. Continents are bordered by shallow continental shelves of varying width and in the North Atlantic; a relatively shallow mid-ocean ridge system divides the abyssal areas into east and west basins. The result is a somewhat irregular rotary wave that has settled into equilibrium with the actual basin topography that exists today (paleoceanographers puzzle over what these systems may have looked like in the distant geological past since continents and ocean basins never stay put!).

Today networks of interconnected rotary waves of varying size are known to operate in the world's oceans, seas, and gulfs. Their intricate patterns of co-tidal and co-range lines have been developed in considerable detail by numerical tide models and, most recently, by means of satellite altimetry. However, the use of co-tidal and co-range lines goes back to a much earlier time.

Among the first to describe tidal behavior in this way was William Whewell who developed co-tidal lines for Great Britain and the coasts of Europe starting in 1833. In 1904, R.A. Harris of the United States Coast and Geodetic Survey used tidal measurements and ocean bathymetry from around the world to construct and publish co-tidal charts for the lunar semidiurnal tide. Harris's methods were not unlike the simple basin-fitting exercise that we have examined in this chapter with the aid of equations (3.6) and (3.7) and it was he who introduced the term **amphidromic system** (from the Greek words *amphi – around* and *dromos – running*). Among the series of co-tidal maps that Harris developed for the world oceans, his map of the amphidromic system for the North Atlantic still compares well today, a century later. Presented in Fig. 3.12, it shows a central amphidromic point with a set of co-tidal lines moving counter-clockwise around it in combination with a second, partially complete counter-clockwise wave centered near the leeward islands separating the Atlantic ocean and the Eastern Caribbean sea. Visually comparing Fig. 3.12 with the latest co-tidal maps for the North Atlantic one notes the location of the main amphidromic point farther to the north and perhaps a clearer indication of the second amphidromic point north of the Anegada passage between the Virgin Islands and Anguilla in the Lesser Antilles.

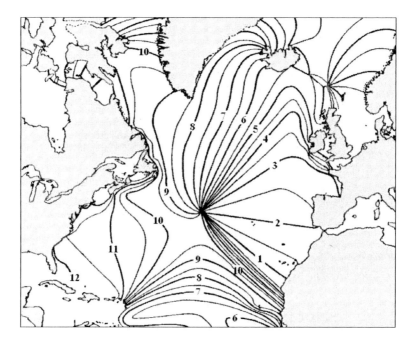

Fig. 3.12. Co-tidal lines for the North Atlantic ocean, after Harris (1904).

Compared to Fig. 3.11, which represents a simple rotating inclined plane in a cylindrical basin, the lines in Fig. 3.12 are neither straight nor evenly spaced but they still show the counter-clockwise rotary motion that operates throughout much of the North Atlantic. The hours identifying co-tidal lines and the position of high water at these times are **Greenwich hour intervals** – lunar hours elapsed since the moon's passage over the meridian at Greenwich, England, or its antimeridian, the International Date Line.

Dynamic tides - Tidal dynamics, the swirl of rotary waves producing the rise and fall of tides as well as the flood and ebb of current flows throughout the world's oceans and seas, unfolds a complex tale indeed. Enough perhaps to spark a bit of wistful envy for the simpler concept of the equilibrium tide calling for nothing more than a pair of static tidal bulges surrounding a spinning earth. Yet a spinning earth whose ocean basins are filled with rotating waves is what we've got. Many of them we would also call trapped waves – moving waves that are hemmed in by the surrounding coastline. However, as with all oceans, the North Atlantic's coastline is not continuous. There are breaks to the north (Labrador, Greenland, and Norwegian seas), south (South Atlantic ocean) and southwest (Caribbean sea and the Gulf of Mexico). At these boundaries the trapping effect gives way and some 'leakage' of the tide wave occurs.

This is where the dynamic part really comes into play. Tides in the Eastern Caribbean sea for example, have a very small range (less than 25 cm in most places) and are predominantly *diurnal* – only one high tide and one low tide occur per lunar day. But the Caribbean basin is too small and too deep to have a natural period of one day or even half a day. Most of its tide comes instead from the nearby Atlantic Ocean where, as noted above, there is an amphidromic point near the entrance to the Eastern Caribbean. Since the semidiurnal (twice-daily) tides have near-zero range close to this point, it is not surprising that very little of this tide finds its way through the island passages into the Caribbean basin. Fig. 3.12 hints at some areas where the lesser of the two rotary waves in the North Atlantic appears to give way to progressive wave motion. For example, the southernmost co-tidal lines in the figure (hour 6 through hour 9) hint at a progressive wave moving from the South Atlantic across the equator into the North Atlantic; the westernmost co-tidal lines (hour 10 through hour 12) suggest a progressive wave moving directly toward the coastline of the Southeastern United States.

Tidal dynamics, in short, are too complex to describe spatially without using computer-driven mathematical models requiring reams of input information, starting with thousands of water depths arranged in fine grids. But a simple fact remains: for all practical purposes, the tide at any point on the watery earth is *steady*. The same rhythm is repeated over and over again as a set of superposed waves, each with its own distinctive period.

3.9 FROM ROTARY WAVES TO SINE WAVES

Fig. 3.6 depicts a plane rotary wave swirling within a teacup. Unlike the rotary waves seen in some textbooks – wavy discs with a 'bump' along the edge to show where high water occurs - this one simply has a tilted plane for its rotating fluid surface. Can a plane surface produce a sinusoidal waveform representing the tide's highs and lows sweeping past a fixed point along the shore? Or do we need a rotating surface with bumps in it? The answer can be worked out mathematically using simple plane trigonometry but a graphical solution serves just as well.

Suppose that the teacup in Fig. 3.6 is a cylinder with a miniature tide gauge attached to the inside wall to measure (sorry) 'tea level'. What will the resulting tide record look like as the tilted plane swirls anticlockwise around the inside of the cup with a period of revolution, *T*? The parts to this puzzle are shown graphically in Fig. 3.13.

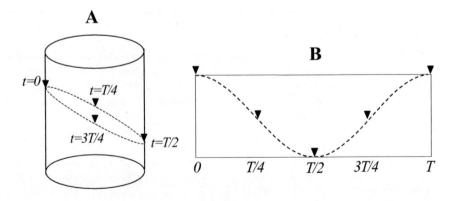

Fig. 3.13. A. Elliptical intersection of a cylinder and a plane; B. Cosine wave, cylinder unrolled.

This is one of those relativistic situations in which we could as easily imagine the tide gauge and the wall moving clockwise while the fluid and the fluid surface remain at rest. In either case the recorded tide, marked, say, on a thin paper cylinder placed inside the cup, would coincide with the intersection of the plane and the cylinder. The intersection of a tilted plane and a cylinder is, of course, an *ellipse* as shown by the dashed line in Fig. 3.13A. If we cut the paper cylinder along a vertical line at $t=0$, unrolling it and laying it flat, we would get a *cosine wave* in the form of the dashed curve in Fig. 3.13B. Need more convincing? Try the reverse - plot a single cosine wave on a piece of paper and roll it up so that the beginning and end of the wave are matched. Your new cylinder should have a tilted plane exactly like the one in Fig. 3.13A - like a sausage cut at an angle with a sharp knife.

As this exercise demonstrates, a plane rotary wave has no difficulty producing the familiar sinusoidal waveform that typifies ordinary tide records at stations along the edge of an ocean basin. To the observer stationed along the shore, it's no longer a rotary wave but a simple progressive wave that passes by.

More about Kelvin waves – The tilted-plane example just presented is intended to give a basic understanding of how sinusoidal tides arise dynamically along the shoreline of an idealized ocean basin with vertical walls. The relationship between rotary wave motion and tides in real ocean basins is considerably more complex as is the theoretical description of a Kelvin wave. The mathematical basis for these waves rests with the assumption that Coriolis force, acting to the right of a tidal stream in the northern hemisphere, is balanced by a hydrostatic force arising from water raised against a shore or other lateral boundary (trapped wave effect). As a result, the Kelvin wave produces no tidal streams normal to the direction of travel, exhibits an exponential decrease in tidal range away from the boundary, propagates as a shallow-water progressive wave traveling parallel to the boundary, and turns counter-clockwise in basins in the northern hemisphere (clockwise in the southern hemisphere). However, amphidromic systems are found that rotate in opposition to the above rule in locations where a physical boundary is lacking or poorly defined. An example of such contrary motion appears in the northernmost of two amphidromic systems in the South Atlantic ocean, suggesting that an unbounded or partially–bounded wave of this type is not a true Kelvin wave. The interaction of multiple rotary waves in complex basins may simply overwhelm this distinction.

4

Harmonic constituents: building blocks of the tide

Tides and tidal currents continually change from place to place. To explore these spatial changes in detail, today's marine scientists use numerical models to create a 'virtual' ocean, bay, or estuary within a carefully defined region, the model domain. But numerical modeling requires a fast computer and a lot of work so it is fair to ask the following question: Is there a simpler way to predict tides and currents through time at some particular place that we're interested in? There is, thanks to a combination of Newton's static theory of the equilibrium tide and the dynamic theory of tides briefly outlined in Chapter 3. The French mathematician and scientist, Pierre-Simon Laplace, is given credit for first developing a dynamic tide theory a century after Newton's time.

Newton's theory recognized that the sun and moon in their 'apparent' motions around the earth set up differentiable forces that produce differentiable tides, or *partial tides*. The dynamic theory of Laplace viewed these partial tides as nothing more than a collection of simple sinusoidal waves, each corresponding to a type of lunar or solar tractive forcing and, in sum, constituting the observed tide. Each partial tide in this mix is identified by a distinctive period of oscillation. England's Lord Kelvin (1824-1907) is credited with devising the **harmonic method of tidal analysis** in which partial tides are extracted from local tidal observations as **harmonic tidal constituents** (details of this method are given in Chapter 10). Theoretically, there can be a large number of these constituents. We have to find the ones most important in terms of their effectiveness in a harmonic model of the tide at any one location.

4.1 HARMONIC TIDAL CONSTITUENTS

Harmonic motion means periodic motion that can be made simple, like the motion of a clock's pendulum, or complex as a result of combining one or more simple motions with different periods. A tidal constituent qualifies as one of these simple motions. In addition to being a long wave with a fixed **tidal period** measured in hours and a wavelength measured in hundreds of kilometers, a tidal constituent also has *amplitude* (feet or meters) and a *phase* (hours or degrees), both of which can vary from place to place. Harmonic analysis determines the amplitude and phase of the tidal constituents. It is done *locally,* one place at a time, so we don't have to worry about wavelength in this case. A graphic example illustrating the general wave characteristics that matter here – *period, amplitude, and phase* - is presented in Fig. 4.1.

The tidal component shown in Fig. 4.1 looks like a tide curve of the type commonly appearing in tidal prediction tables or nautical guides. It isn't quite. As the name implies, it is only one constituent out of several that must be combined to make an accurate tidal prediction. The curve shown in Fig. 4.1 is a simple *cosine wave* and this particular one has a period of 12 hours and 25 minutes (12.42 mean solar hours). This identifies it as the *main lunar semidiurnal* or **M_2 tidal constituent**. The symbol 'M_2' is a kind of shorthand for 'moon' and 'twice-daily'. Except in places where the type of tide is predominantly *diurnal* (one high and one low per day), M_2 is the dominant tidal constituent in terms of amplitude. If a cosine wave with a period of exactly 12 hours (12.00 mean solar hours) had been shown instead in Fig. 4.1, it would have been called the *main solar semidiurnal* or **S_2 tidal constituent**. Other constituents are defined in a similar way but before we go into that, let's see what happens when we combine (superpose) a pair of tidal constituents like M_2 and S_2.

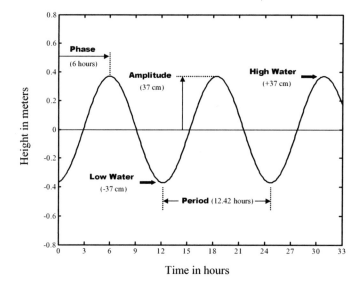

Fig. 4.1. The M_2 tidal constituent with period of 12.42 hours.

Combining tidal constituents – The government agency responsible for tidal predictions in the United States is the **National Ocean Service (NOS)**, a branch of the National Oceanic and Atmospheric Administration (NOAA). A harmonic analysis of water levels conducted by the NOS at Sewells Point in Hampton Roads, Virginia determined the M_2 and S_2 constituent amplitudes at this location to be 36.6 cm and 6.4 cm, respectively. Fig. 4.2 shows how two cosine waves with these amplitudes might look using the same plotting scale as Fig. 4.1 and giving both waves a phase of 6 hours. We're not being very specific about the actual starting time just yet but exactly six hours after startup, the combined height that results from adding the M_2 wave to the S_2 wave is 36.6 + 6.4 = 43.0 cm. If we do a similar addition of corresponding points on the two curves at all other times, we end up tracing the solid curve in Fig. 4.2 as the combination of M_2 and S_2. As you have probably already guessed, the greater range of heights produced by this particular combination would be called a *spring tide*, a condition that applies when the M_2 and S_2 waves are *in phase* or nearly so.

Although the M_2 and S_2 tidal constituents are exactly in phase at hour 6 in Fig. 4.2, you can see that soon afterwards their phases are no longer the same: S_2 has reached its third high water at hour 30 but M_2 has almost another hour (about 50 minutes) to go before its third high water occurs.

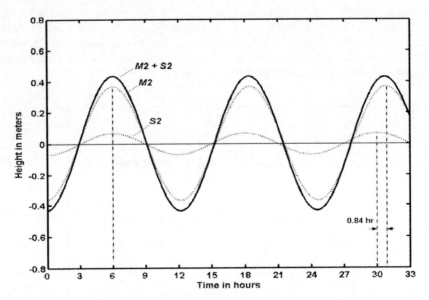

Fig. 4.2. Spring tide with the M_2 and S_2 tidal constituents in phase.

To compare the change in phase of two tidal constituents, we will refer to the constituent *speed* expressed in degrees per hour and we'll need to use at least six significant digits to represent them with sufficient accuracy in tidal calculations. The M_2 constituent speed, for example, is 28.9841 degrees per mean solar hour and the S_2 constituent speed is 30.0000 degrees per mean solar hour, numbers obtained by dividing 360 degrees by the period of the constituent in hours. In terms of constituent speed, S_2 is obviously 'faster' than M_2 in the sense that it completes more cycles in a given amount of time.

Now take a look at Fig. 4.3 below. After 180 hours, S_2 has clocked 15 complete cycles compared to about 14.5 cycles for M_2, and the two constituents are now almost completely out of phase[1]: M_2 is high while S_2 is low. Combining them now (solid curve in Fig. 4.3) gives us a *neap tide* with lower highs and higher lows – the minimum tidal range produced by these two tidal constituents. Note that the sum $M_2 + S_2$ in Fig. 4.3 indicates that we are still combining tidal constituents by adding them, recognizing that one (S_2) has a negative height at hour 180 which is subtracted from the positive height of the other constituent (M_2) at that time.

---

[1] More precisely, M_2 and S_2 reach *quadrature* after 177 hours, one-quarter of a lunar month of 29.53 days marking the recurrence of lunar phases.

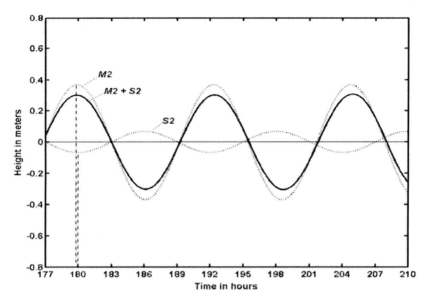

Fig. 4.3. Neap tide with the M_2 and S_2 tidal constituents out of phase.

Switching to a time scale of days instead of hours, Fig. 4.4 shows how the combination of M_2 and S_2 produces the same 14.76-day *spring-neap cycle* described in Chapter 2 using the equilibrium tide concept, only this time we're using a pair of simple cosine waves with no need for tidal bulges.

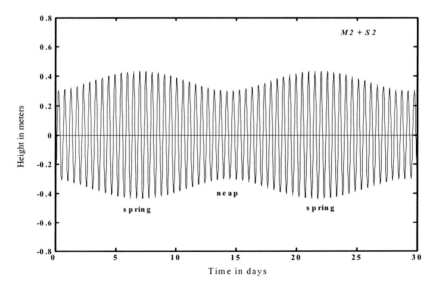

Fig. 4.4. Spring-neap cycle produced by the M_2 and S_2 tidal constituents.

And that's just the beginning. The next constituent that appears at this stage of the harmonic model building process is the *larger lunar elliptic semidiurnal* or **N_2 tidal constituent**. It has a fixed period of 12.66 hours (speed = 28.4397 deg·hr^{-1}), which seems routine until a cosine wave with this period is combined with M_2. Then another cycle very similar to the spring-neap cycle unfolds – the *perigean-apogean* cycle, which, as pointed out in Chapter 2 (Sec. 2.5, p.16) occurs because of the moon's elliptical orbit and varying distance from the earth during the elliptic month of 27.55 days.

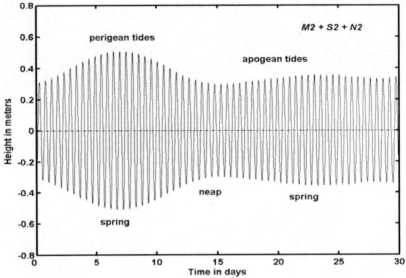

Fig. 4.5. Tidal cycles produced by the M_2, S_2, and N_2 tidal constituents.

4.2 THE HARMONIC MODEL OF THE TIDE

After adding just two more tidal constituents we will have a **harmonic model** for tidal predictions that gives a fair representation of the **astronomical tide** caused by the moon and sun at many worldwide locations including Chesapeake Bay. You will recall from Chapter 2 that the moon in its orbit around the earth reaches a maximum declination north or south of the equator every 13.66 days and that during these times *tropic tides* occur (tides that display a diurnal inequality or difference in successive high and/or low water heights each day). In between these times, when the moon is over the equator, *equatorial tides* occur with little or no diurnal inequality.

To simulate the 'on-off' behavior of the *tropic-equatorial cycle*, two tidal constituents rather than one are needed. These are the *lunar-solar declinational diurnal* or **K_1 constituent** (period = 23.93 hours) and the *lunar declinational diurnal* or **O_1 constituent** (period = 25.82 hours). Note that the constituent periods are different, which means that their waveforms will pass in and out of phase with time. K_1 and O_1 have roughly similar amplitudes and work as a pair so that the diurnal tide is greatest when the two are in phase (tropic tides) and least, or near zero, when out of phase (equatorial tides). Adding K_1 *and* O_1 to the mix with their NOS derived amplitudes for Sewells Point, we obtain the predicted tides shown in Fig. 4.6. Here the diurnal inequality that appears during tropic tides shows up more in the high waters than in the lows. This is typical of tropic tides throughout the Chesapeake Bay region.

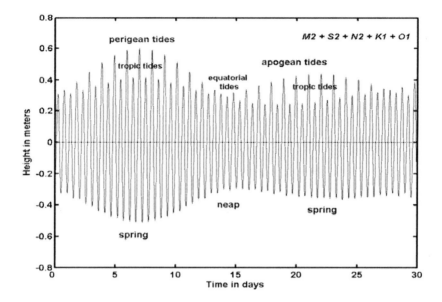

Fig. 4.6. Tidal cycles produced by the M_2, S_2, N_2, K_1, and O_1 tidal constituents.

The predictions in Fig. 4.6 illustrate the general appearance of the types of tide that can result after adding together the five tidal constituents discussed above. However, the tide curves shown are unlikely to represent an actual record for any given period of time because, although the actual tidal amplitudes at Sewells Point were used, the phases of the five constituents were all given the same value for purposes of illustration. In reality, the five constituent phases are normally not the same and tidal prediction curves can be quite different depending on the combination of numbers that actually applies. In fact, the phase combination has to be just right or the times of the predicted tides, the highs and lows in particular, will be off. That can spell disaster for a ship entering a waterway with less water under her keel than expected, thanks to a bad tidal prediction. If the height of the predicted tide is approximately correct but predicted to happen at the wrong time, we're most likely looking at a phase error.

4.3 TIDAL TYPES AND THE TIDAL FORM NUMBER

Although amplitude and phase are equally important in generating accurate tidal predictions, tidal constituent amplitudes determine the *type* of tide (diurnal, semidiurnal, or some combination of the two). The **form number (F)** is a convenient way to define tidal type. It is computed as the sum of the main two diurnal amplitudes divided by the sum of the main two semidiurnal amplitudes using the formula below:

$$\text{Tidal form number: } F = \frac{K_1 + O_1}{M_2 + S_2}$$

When the form number is less than 0.25, we have *semidiurnal* tides – as in Chesapeake Bay and along most of the U.S. East Coast (the form number at Sewells Point is 0.21). Between 0.25 and 1.5, tides are *mixed, predominately semidiurnal* (U.S. West Coast) and between 1.5 and 3.0, *mixed, predominately diurnal* (Manila, Philippines). Above 3.0, the tidal form is fully *diurnal* (Pensacola, Florida). Tidal types are summarized in Table 4.1.

Table 4.1. Tidal types defined by the tidal form number.

Tidal Type	Form Number	Typical Form
Semidiurnal Tides	Less than 0.25	
Mixed, Semidiurnal	0.25 – 1.5	
Mixed, Diurnal	1.5 – 3.0	
Diurnal Tides	More than 3.0	

A complete harmonic analysis of the tide at Sewells Point and elsewhere in Chesapeake Bay would include many other tidal constituents whose origins are more complex than the primary five. But while these additional tidal constituents increase the final accuracy of a given day's predictions, it is worth noting that M_2, S_2, N_2, K_1 and O_1 by themselves usually account for more than three-quarters of the normal daily variation in water level at Chesapeake Bay tide stations. We can certainly add other constituents but skill is needed to pick the right ones for a given area. Even if we pick the right ones, but give them the wrong amplitude or phase, we add disinformation to the prediction. And much of the remaining percentage is due to non-tidal variations caused by weather - winds and changes in atmospheric pressure. We'll take a look at some interesting examples of weather-induced change in Chapter 7.

4.4 REFERENCING TIME AND HEIGHT OF TIDE

The tide curves presented in Figs. 4.1 through 4.6 show the form of the tide well enough but otherwise they're not very useful for absolute measures because their scales have no reference marks – they are 'floating' in time and space. The way these figures are presented, we don't really know the actual times and heights of the tide, just that they're so many hours after and so many meters above or below 'zero', whatever that is. To give practical meaning to a tidal prediction, carefully defined references for time and height are required. Referencing time is fairly straightforward but referencing height is less so. A separate chapter (Chapter 5) will be devoted to the task of introducing a key reference, the **tidal datum**, and explaining how it is used to get a firm handle on referenced water levels.

Time – In most tide tables, time is displayed in hours after the 'zero' hour starting at midnight at the beginning of each day using some form of **standard time**. Everyone's familiar with the standard time zones on land in countries around the world, as well as certain seasonal conversions such as daylight savings time. These are legal definitions of time but they can trip us up unless we remain aware that **universal time** is the underlying standard. Universal time is based on the average apparent motion of the sun relative to a longitudinal reference called a *time meridian*. *Greenwich Mean Time* (GMT) results when the prime meridian at Greenwich, England, is used; it is exactly noon or 1200 GMT when the *fictitious mean sun* is directly over the Greenwich meridian. Why fictitious? Because, in its apparent motion around the earth, the real sun appears to speed up and slow down at different times of the year. You wouldn't want a watch that ran like that; neither do the experts who keep track of solar time. Lacking a smoothly running 'average' sun, they've created one instead.

Twenty-three additional time meridians are located at 15-degree intervals east and west of the one at Greenwich. To set up time zones, a 15-degree interval of longitude is used. The U.S. East Coast falls within the interval centered on the time meridian at 75 degrees west longitude (75°W). Since the mean sun moves west at a constant 15 degrees per hour, it passes over this meridian at noon local time exactly 5 hours after Greenwich noon (1200 GMT +5 hours). The same *Universal Standard Time* (GMT minus 5 hours), also known as *Local Standard Time* (LST), applies to all points in the zone between two boundary meridians located 7½ degrees to either side of 75°W.

Unless you're a ship's navigator or live in the United Kingdom, Iceland, Portugal or West Africa, it's unlikely that your timepiece is set to GMT. But in eastern North America, for example, official tide tables such as those produced by the US NOS refer to the time meridian of 75°W and make no mention of Eastern Standard Time or Eastern Daylight Time (EDT). It's up to you to equate LST 75°W with whatever time you have on your watch. If it's Eastern Daylight Time, the times of high and low waters apply an hour later than the times published in the NOS tide tables.

Did that last statement make you pause for a second? If so, look at the tide curve in Fig. 4.2 and imagine that the time scale showing the first high tide at 6 hours is actually a ruler (one marked in hours instead of centimeters). Assuming this is a prediction made using Local Standard Time, you would shift the ruler to the left one hour to change from standard time to daylight time. Your ruler, like your watch, would then place the same high tide at 7 hours, one hour later than the original high water prediction in local standard time. Though logical, it's still easy to make a mistake that results in the wrong time being marked down for either a prediction or an observation.

Tidal height and water level – Tidal height is the water's displacement measured vertically, up or down, from still water - the water level presumed to exist in the absence of a tide. The word *height* is normally used when referring to tidal height predictions but *water level*[2] is the term used in the United States when referring to measurements of the water's elevation above a recognizable point on land. For example, if the television news team reports that the tides during a 'northeaster' will be a foot higher than normal, that may turn out to be an accurate prediction but it assumes we know where 'on earth' normal is. We really have no clue unless we have access to a reference water level, the tidal datum. The most common example of a tidal datum is **Mean Sea Level (*MSL*)**, a long-term average of water level determined from observations at tide stations. Other examples include **Mean High Water (*MHW*)** and **Mean Low Water (*MLW*)**. A slightly lower datum called **Mean Lower Low Water (*MLLW*)** is used as the reference for soundings on all US nautical charts. An analogous datum, **Lowest Astronomical Tide (*LAT*)** is used in the United Kingdom and elsewhere. These are important and rather carefully defined water levels that take considerable time and effort to establish and maintain; more will be said about them in Chapter 5.

Recorded tides – Tidal observations are the basic water level readings that come from a *tide gauge*, an instrument for measuring water levels at a fixed time interval (e.g., a tenth of an hour) and determining recognizable daily extremes such as the ones shown in Fig. 4.7.

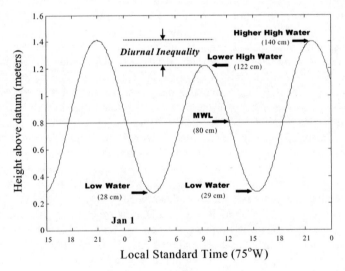

Fig. 4.7. Tide curve illustrating recorded high and low waters above a tidal datum.

The 'zero' level used to reference water level as it is being recorded is called the **station datum**. The location of the station datum is arbitrary; it doesn't refer to a specific physical boundary, such as the sea floor lying below the tide gauge, rather its

---

[2] The term *water level* is used in the US where there are measurable tides in the Great Lakes; *sea level* is the preferred term in the United Kingdom and most other countries.

key attribute lies in remaining fixed relative to the land. It's fixed when the tide gauge is installed (usually on the end of a pier) and care is taken to see that it doesn't change or 'float' while the gauge remains in operation. That way, all water levels placed in the station's tidal database refer to the same physical point.

What measures typically go into a tidal database? First of all, the extreme tides that occur each day are measured in feet or centimeters and recorded separately. For example, two low waters (28 cm and 29 cm) and two high waters (122 cm and 140 cm) are marked for January 1 in Fig. 4.7. Note that there is a greater *diurnal inequality* in the high waters (18 cm) than in the low waters (1 cm), a typical scenario in Chesapeake Bay during tropic tides. In addition to the daily extremes, water levels are recorded each hour on the hour. For the 24-hour period on January 1, the hourly values for the day yield a mean water level of 80 cm. Daily mean and monthly mean water levels are short-term averages not to be confused with *MSL*.

After glancing at the diurnal inequality in Fig. 4.7 and noting that it mostly applies to the high waters, not the low waters, you may wonder why a separate recording of lower low water is something anyone would want to put in the database for the Chesapeake Bay region. The reason is that they are needed to define *MLLW*, the nautical chart datum in use throughout the United States and its territories. Although the difference between *MLW* and *MLLW* isn't very large in Chesapeake Bay, it is much larger on the west and gulf coasts of the United States where tides are mixed or of the diurnal type.

Tidal currents - At first glance, tidal currents would appear to need an entirely different type of representation than tidal heights. Tidal height is a 'scalar quantity' because only one number (height) is needed to describe it. Current, on the other hand, is a 'vector-quantity' requiring two numbers (speed and direction) to describe it. Fortunately, there are ways to convert the tidal current at a given location (and depth) to a scalar quantity that can be analyzed harmonically the same way tidal heights are analyzed. Nature sometimes does the conversion for us; e.g., in narrow river channels where the current can only flow in two directions: upstream or downstream (landward or seaward). By convention, the upstream tidal current in such channels is called *flood current* while the downstream current is called *ebb current*. In less restricted waterways, such as a channel in the open bay, the current direction may change continually over time but it will usually display two dominant directions in which the largest current vectors (the longer ones symbolizing the greatest speed) are aimed. Although it's not always possible to discriminate between landward and seaward, usually we can do so by looking at the coastal configuration and then labeling one direction 'flood' and the other 'ebb'. The important thing is to realize that such currents flow mostly in one direction for half a tidal cycle, then reverse to flow mostly in the opposite direction the other half, doing it over and over again.

What can we do with these reversing flood and ebb currents? We can plot them as shown in Fig. 4.8 where they look very much like a plot of tidal heights. All that is required is to plot current speeds above the zero line in the dominant flood direction using positive numbers and below the line in the dominant ebb direction using negative numbers. Nice and simple but there's a slight catch. In a bi-directional plot like the one in Fig. 4.8, zero current or *slack water* is implicit each time the current reverses. But even in an estuary, currents at any given point are seldom completely bi-directional, so *minimum current* may be a more accurate term to use than slack. However, the plot itself in Fig. 4.8 is not inaccurate. It simply represents the part of the flow (the vector component) aligned with the **principal axis** defining flood and ebb directions.

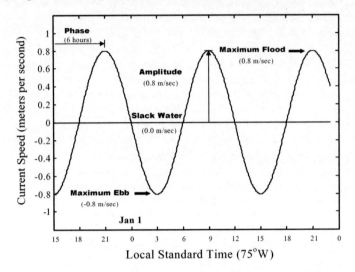

Fig. 4.8. A plot of tidal current in the flood and ebb directions.

Harmonic analysis can be applied to a tidal current record of the type shown in Fig. 4.8 using the same methods applied to tidal heights. The same tidal constituents including M_2, S_2, N_2, K_1, and O_1 will appear except that the amplitudes will be given in cm/sec instead of cm. Although the 'zero' on the current scale has a physical basis (slack water and the reversal of flow direction) it, too, is not an absolute reference. Tidally averaged currents (currents that are presumed to exist in the absence of a tide) are especially common in tidal rivers where the mean current at the surface usually has a negative value indicating a net ebb flow. In fact, far upstream near the head of tide, the actual current during times of high river inflow may not reverse at all but simply show a sinusoidal variation in the strength of the ebb current.

4.5 SEASONAL TIDES AND SHALLOW WATER TIDES

The origin of the **seasonal constituents**, Sa and Ssa, was described in Chapter 2 (Sec. 2.8). Although their amplitudes are largely dependent on non-tidal forcing, they can be added to the mix of tidal constituents because their periods conform to two of the astronomical periods: one year for Sa and one-half year for Ssa. Ways of including these long period constituents along with those obtained from a short series analysis (i.e., 29 days) are presented in Chapter 8. At the other end of the scale, tides with periods shorter than the semidiurnal tides arise in shallow water areas - the **overtides**.

What's an overtide? – A sound wave with a single fundamental frequency appears 'flat' to the ear but it can be made more vibrant by adding higher harmonics that are multiples of the fundamental frequency. Musicians call them 'overtones'. It should not be surprising to learn that special tide waves called overtides exist at certain places in the coastal ocean. The M_2 constituent for both tides and currents can give rise to a series of new constituents called M_4, M_6, and M_8 with exactly two, three, and four times the fundamental M_2 frequency, respectively.

Overtides exist for the S_2 tidal constituent as well as others, but, as a rule, the amplitude of each overtide is much smaller than its parent and becomes smaller still as the frequency increases so that only a few overtides are ever considered significant. Moreover an overtide, like an overtone, doesn't just arise all by itself. A cello produces its rich sound not by accident but through the artful finger work of the cellist. You could say that a very shallow or flow-constricted area, such as a broad shoal or a tidal inlet, does some artful work on a progressive tide wave that has the effect of restricting the trough while advancing the crest. The result is a set of overtides or *shallow water tides* with a unique and interesting property – because their frequencies are exact multiples of a fundamental frequency, they are 'phase-locked' with themselves and their parent wave. This means they cannot pass in and out of phase with time, as do M_2 and S_2 during the spring-neap cycle. Instead, the shape of the fundamental waveform is permanently distorted so that, for example, the average duration of a rising tide may become shorter than the average duration of a falling tide. The same *asymmetry* can be even more pronounced in the flood and ebb durations of tidal currents.

What's a Kelvin machine? - Once the tidal constituent amplitudes and phases are known, tidal current predictions can be made in the same way that tidal height predictions are made. Today, both kinds of predictions are easily done with a desktop computer. From Lord Kelvin's time until the early sixties, they were done by machines that worked on the principal of the simple machine sketched in Fig. 4.9.

As represented schematically in Fig. 4.9, the basic mechanism driving a Kelvin machine is a motor-driven wheel that converts rotary motion into linear up-and-down motion – just the reverse of what the reciprocating engine does in a car. The period of motion is determined by the rotary speed (rpm's) of the wheel controlled through gears connecting the wheel to the motor. Amplitude *a* is determined by the center offset distance, ½ *a*, of a round pin connected to the wheel and extending through the slot in the inverted 'T-bar' shown in the figure.

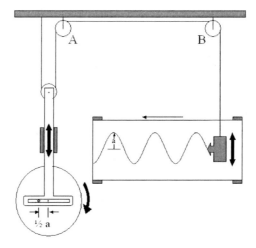

Fig. 4.9. A Kelvin machine predicting a single tidal constituent.

Unlike the crankshaft in a car engine, the slotted T-bar moves up and down through guides so that it always remains vertical. Phase is set by unlocking the gears and turning the wheel to the appropriate starting position before re-locking to start. A pulley-guided wire transfers all of the linear motion just described to a weighted pen shown on the right (the moving pulley imparts a mechanical advantage so that the pen's linear displacement is always twice that of the T-bar). The pen, in turn, draws a cosine wave with the desired characteristics on a moving roll of paper – a tidal prediction for a given place and time. The system illustrated in Fig. 4.9 would, of course, generate only a single cosine wave such as might be used to represent the M_2 tide. To add other tidal constituents to the predictions, other T-bar units are attached to the machine and linked to the moving wire by additional sets of pulleys arranged between points A and B.

4.6 SIMULATING TIDES WITH A SERIES OF COSINE WAVES

Tidal constituents are represented mathematically by one of the simplest functions from basic trigonometry – the **cosine wave**. Getting the numbers together to compute and draw a cosine wave used to be a lot of trouble in the days when everyone had to look them up in a table, going in with the argument in degrees or radians (360° =2π radians) and coming out with the cosine value, a fractional number between –1 and 1. Now if you want to find the cosine of 60° you simply enter the number '60' on your hand-held calculator, punch the button labeled 'cos', and read the number '0.5' in the answer window. Note that a *sine wave* looks exactly like a cosine wave except for its phase, which is shifted relative to that of the cosine wave by a quarter cycle.

The math needed to represent a tidal constituent is also very simple. Since the input value (e.g., the x-axis in Fig. 4.1) is the time, t, given in hours, we need a conversion factor to change it to degrees. This factor is simply the frequency or *speed* of the wave in degrees per hour. We denote the speed by the symbol ω, where $\omega = 360°/T$ and T is the period in hours. The argument for the cosine function used to draw the curve in Fig. 4.1 will then be

$$\omega_1 (t - t_1) = \frac{360^0}{T} (t - t_1)$$

where ω_1 is the speed of the M_2 tidal constituent, T is its period, and t_1 is its phase in hours. The phase argument can also be expressed in degrees using $\phi_1 = \omega_1 t_1$. The complete formula needed to predict the tidal height, h, using the M_2 tidal constituent is then

$$h = h_0 + R_1 \cos(\omega_1 t - \phi_1) \tag{4.1}$$

where h_0 is the reference water level relative to mean sea level, and R_1 is the symbol representing the M_2 wave amplitude for a given year (remember that this amplitude varies slightly from one year to the next during the 18.6-year lunar cycle). We could choose mean sea level itself for the reference by setting $h_0 = 0$. Although this would give us negative numbers to deal with about half the time, it is a way of predicting tidal heights above 'normal' or average water level. Using

$$h_0 = MSL\text{-}MHW$$

would predict heights above the 'normal' high water level. If we choose Mean Lower Low Water (*MLLW*) instead for our reference level by setting

$$h_0 = MSL-MLLW$$

the resulting predicted tides will represent heights above MLLW. After choosing MLLW as the reference, all that is necessary to change the depth shown on a U.S. nautical chart to one corrected for the tide is to add the predicted tide, h, to the chart depth.

As noted earlier, the phase of the cosine wave is given by the interval between the starting time (time origin) and the first high water occurrence. According to Newton's equilibrium theory with its twin 'tidal bulges', the phase should be the same as the celestial body producing the high (bulge); i.e., the moon should be passing either above or below the local meridian at the time of high water. In reality, as the dynamic theory shows, the tidal constituent phase is not fixed in this way but, like the tidal constituent amplitude, varies from place to place. Other tidal constituents will have different amplitudes and phases that also vary spatially. The frequencies, of course, are constant with precisely known values such as those given in Table 4.2.

Table 4.2. Frequencies (degrees per hour) of selected diurnal, semidiurnal and quarter-diurnal tidal constituents[3].

K_1	15.0411	O_1	13.9430	P_1	14.9589	S_1	15.0000
M_2	28.9841	S_2	30.0000	N_2	28.4397	K_2	30.0821
M_4	57.9682	S_4	60.0000	MS_4	58.9841	MN_4	57.4238

To get amplitude and phase values for a set of tidal constituents (M_2, S_2, N_2, K_1, O_1, etc.), harmonic analysis is performed on a tidal record, a time series of recorded water levels obtained with a tide gauge. If m tidal constituents are obtained from the analysis, then the following prediction formula can be used:

$$h(t) = h_0 + \sum_{j=1}^{m} R_j \cos(\omega_j t - \phi_j) \quad (4.2)$$

In this equation the summation symbol tells us that we have to add all m cosine terms representing the set of tidal constituents identified by the subscript, j. In adding the m terms together it doesn't matter which one comes first so long as the right combination of amplitude, frequency, and phase is used. For example, if $j = 1$ refers to the M_2 tidal constituent, its amplitude, frequency, and phase would be represented by H_1, ω_1, and ϕ_1, respectively. All we have to do is enter these numbers in a summation formula and plug in a time, t, to calculate the sum of the tidal constituents at that time. Then a new value of t, say thirty minutes later, is plugged in and so on until the desired span of time for developing the tidal predictions is covered. This process goes very quickly using nothing more than a spreadsheet program on your personal computer. How the amplitude and phase values are derived for the tidal constituents listed in Table 4.2 is explained with the help of some interesting examples in Chapter 7.

---

[3] The constituents MS_4 and MN_4 in Table 4.2 are examples of *compound* tides deriving from shallow-water interactions between M_2 and S_2, M_2 and N_2, respectively.

5

Tidal datums: finding the apparent level of the sea

The sea is never still. Wherever there are tides and atmospheric forces that affect the sea, water levels will continually change. In previous chapters, I explained why the change due to the astronomical tide occurs and how to predict it. But I didn't tell you where you could find a predicted tide – a predicted high water, for example – relative to the ground you may be standing on, or the bottom you may be sailing over.

5.1 DEFINITIONS AND USE OF A TIDAL DATUM

To relate water levels to the solid earth, we need a *tidal datum*, a vertical reference for reckoning heights (or depths) that corresponds to a particular phase of the tide – high water, low water, or some other recognizable level. We also need a way to define the datum so that it remains consistent at different times and at different places. To remain consistent over time, a tidal datum is defined by taking an arithmetic average of the water level in question over a number of years. In the United States, for example, the tidal datum of *mean high water* is defined as the average of all the high waters measured at a tide station over a period of 19 years; the average of all low waters in this period defines *mean low water*. The tidal datum of *mean sea level* is defined as the average of all the hourly heights of tide measured over the same period. More will be said about these and other tidal datum definitions in Sec. 5.8 of this chapter.

Apart from their role as a reference for tide predictions, the tidal datums of mean high water, mean sea level, and mean low water have special relevance as *boundary markers*. In the continental United States, the mean high water mark defines the limits of riparian ownership in some coastal states whereas the mean low water mark defines it in others. If you own a home or an outbuilding that falls within certain flood zones established by the Federal Emergency Management Agency (FEMA) in the low-lying coastal areas of the U.S., you probably have federally mandated flood insurance. Those zones are defined by selected elevation contours referenced to mean high water. Federal and state legislation for wetlands preservation and watershed control frequently reference one or the other of these datums. Over the past thirty years, I've received many calls from property owners wanting to know how to locate the mean high water line on their land. The information included at the end of the chapter will show you how.

I used the word 'apparent' in the chapter title because a tidal datum is based entirely on water levels as they appear to an observer at one location. Let's explore that statement through an example. Suppose I'm an observer who, on a certain morning, is watching a rising tide at Gloucester Point on the York River in Virginia. As the water line slowly creeps across the beach, I watch it reach the limit of its landward excursion. If I drive a stake into the ground at that point, I will have found the morning's **high**

water mark, or so it appears to me. But first I telephone a friend farther up the York River at that same moment in the town of West Point. He tells me that the water is still rising on his beach (because of the tidal crest moving up the river as part of a progressive wave). In other words, my friend is oblivious to any clue from me as to the 'when' and 'where' of the corresponding high water mark at West Point. He must find that mark based on his own observation. And assuming he does, there's no reason to expect that his high water mark will have the same *absolute* elevation as mine; i.e., have the same height above a common level plane as determined by a precision leveling survey conducted between Gloucester Point and West Point. As explained in Chapter 6, tide wave hydrodynamic behavior frequently leads to a difference in tidal heights measured from point to point above a level plane.

5.2 ON THE LEVEL? IT'S A GEODETIC DATUM

The survey just referred to is not an ordinary one. Although the towns involved are only thirty miles (48 km) apart, a leveling survey between them must take the curvature of the earth into account in order to compare high water elevations to the nearest centimeter. In that case, it's called a **geodetic survey** and the 'level plane' would be called a **geodetic datum**.

Although both tidal datums and geodetic datums have been used in the past as a type of sea level reference for elevations, the similarity ends there. Geodetic datums are primarily used to reference elevations on land: the height of a mountain peak, for example, above the geodetic datum of mean sea level. Mean sea level in this case is a reference spheroid: a mathematically defined three-dimensional surface that approximates the size and shape of the earth.[1] In North America, this type of surface was initially transferred across the United States and Canada at benchmarks established on the ground through first-order triangulation leveling networks, a task now taken over by earth-orbiting satellites. Implicit in the definition of a reference spheroid is the assumption that gravity acts normal to it at every point on its surface, hence giving the surveyor the ability to transfer geodetically referenced elevations by leveling; i.e., by sighting a surveyor's rod through a telescopic instrument whose horizontal optical axis is 'leveled' or aligned normal to local gravity on a **geopotential surface**.

The geodetic datum presently in use in the United States is the *North American Vertical Datum of 1988* (*NAVD 88*). It replaces the *National Geodetic Vertical Datum of 1929* (*NGVD 1929*), which was originally known (confusingly) as the Sea Level Datum of 1929. The year indicates the time when adjustments were last made between portions of the separate leveling networks stretching across the continent. The point to keep in mind is that the tidal datum of mean sea level differs from geodetic mean sea level at most locations. As will be explained shortly, tidal datums are *dynamic* and *NGVD 1929*, until very recently, was one of the 'static' vertical references against which their temporal change in elevation could be measured.

5.3 A TIDAL DATUM BEGINS WITH A TIDE STAFF

---

[1] At a scale of meters and centimeters, the real earth is not a perfect spheroid but a pear-shaped, lumpy object called the *geoid*. The science of geodesy seeks to better define the geoid's size and shape in deriving a reference for the precise location of points on earth.

All that was previously said about high water marks applies at other places and to other phases of the tide as well – phases such as *low water* or *lower low water* where the tides are semidiurnal or mixed (two low tides of unequal height per lunar day). Driving stakes into the ground is not very practical though, so I may want to use a vertical scale mounted on a piling, a *tide staff*, for marking water level positions. That way, I can record a number of high or low tides in succession and come up with the beginnings of a tidal datum by averaging their heights and marking the average position on the staff. The vertical datum measured on the tide staff can then be transferred to a point (or a contour) on land using a surveyor's rod and level as illustrated in Fig. 5.1.

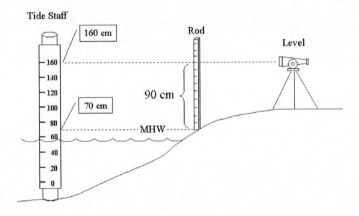

Fig. 5.1. Transferring a high water mark with a rod and level.

Let's say that a tide staff like the one in Fig. 5.1 has been used to visually observe and record a series of consecutive high water heights that are in turn used to calculate an arithmetic mean. We can call the result *Mean High Water (MHW)* but if the series is not a long one, a qualifier is sometime used to state the averaging period; e.g., **Monthly Mean High Water (*MMHW*)** would be a better term for a mean value based on a single month of high water measurements. If an average were computed over 19 years using the highest tide observed each day, we'd call that **Mean Higher High Water (*MHHW*)**. Otherwise, for a single high water average, we'd include both highs that normally occur every day where tides are the semidiurnal type. Remember that lunar tides occur about 50 minutes later each solar day; when the second tide of the day eventually slips past midnight and into the next day, there may be only one high or one low left to record. Of course, for a long series, we wouldn't do the number writing ourselves. We would install a *tide gauge*, an instrument that automatically records tidal heights at regular intervals (e.g., one reading every 6 minutes, or ten times an hour).

Once a datum such as *MHW* has been calculated, we can transfer it to a point on the ground using a surveyor's rod and telescopic level. To see how this works, imagine that a short-term series average has given us a *MHW* elevation of 70 cm on the tide staff shown in Fig. 5.1. After setting up the level, a surveyor looking through the telescopic sight to the tide staff will see a horizontal line in the eyepiece intersecting the graduated scale of the staff (or the extended rod placed against the staff with its base at the staff zero mark). Recording a reading of, say, 160 cm, the surveyor knows that the height of the instrument's optical axis is 160 cm above the staff zero mark. Next the surveyor's assistant holds the rod vertically upright in the line of sight and moves it in small steps across the shore toward the leveling instrument. The surveyor signals him to stop when a rod reading of 90 cm appears in the level (i.e., 160-70 cm). At this point, the base of the rod will mark the *MHW* elevation (70 cm) on the ground. By progressing along the shore in similar fashion, the *MHW* contour can be developed.

5.4 AND ENDS WITH A TIDAL BENCH MARK

Or hopefully so. As you can see, setting up even a short-term datum requires a bit of work. It would therefore be a shame if something or someone came along and destroyed the tide staff, leaving us with numbers but nothing to connect them to! **Tidal bench marks** are installed to make certain that doesn't happen.

A benchmark can be any solid object fixed in its position on land - a metal rod driven into the ground, for instance, or a brass disk fixed to a rod set in concrete. To ensure datum preservation, it's a good idea to place several tidal benchmarks a short distance inland above the highest water levels likely to occur. After that, another leveling survey is undertaken to determine the elevation of each benchmark above the station datum (tide staff zero). The resulting information established for the benchmark, either imprinted on the mark or recorded elsewhere, is the elevation of the mark above the tidal datums subsequently established as well as the station datum. Now the loss of the tide staff wouldn't be a total disaster. Once the necessary tidal observations have been made, *MHW* and all other datums resulting from the measurements will continue to reference their height to tide staff zero via the benchmarks, even if the staff is no longer there.

Considering the tide cycles described in Chapter 4 that we always have, in addition to coastal storms and other transient events that routinely affect water levels over a period of several days, it is easy to see that one month of tidal measurements is pretty much the bare minimum for averaging out these variations and securing a partially stable datum. However, as stated earlier, a 19-year record is required to determine a full-fledged tidal datum for official use in the United States and its territories.

5.5 TIDAL EPOCHS: WHY 19 YEARS?

If, instead of tidal heights, we were to consider fluctuations in the price of electricity or heating oil, we'd take the complete seasonal cycle into account before comparing the average cost of home heating from one year to the next – it wouldn't do to leave out winter prices, for instance. So if an ultra-long cycle of variation is known to exist, we would need to include all of it in the averaging process to 'fix' the level of the sea. The longest cycle having practical significance is the **19-year Metonic cycle** discovered by Meton, a fifth century Athenian astronomer. It spans the interval required for new and full moon to recur on the same day of the year. However, Meton's cycle is not as important in terms of its influence on the astronomical tide as the 18.6-year precession of the moon's nodes described earlier in Chapter 2. The lunar node cycle affects the range of the tide from one year to the next because it involves a slow, periodic change in the inclination of the moon's orbit relative to the earth's equatorial plane: effectively a change in the limits of the moon's maximum declination north and south of the equator, a feature that modulates the monthly tropic-equatorial tides (see Fig. 2.13).

The moon, however, is not the only factor here. The sun goes through its own declinational cycle producing the *Solar Annual* (Sa) and *Solar Semiannual* (Ssa) tidal constituents as described in Chapter 4 (Sec. 4.5). As everyone knows, global weather is intimately geared to this winter-summer declinational cycle and adds its own unique contribution to tidal variations during the year. In order to cover a whole number of annual cycles, and just a little bit more than a full 18.6-year lunar node cycle, a period of 19 years is traditionally used to determine the principal tidal datums at government tide stations in the United States. The 19-year period is known as a **tidal epoch**.

5.6 TIDAL DATUMS IN USE AROUND THE WORLD

The principal tidal datums used in the United States include Mean High Water (*MHW*), Mean Low Water (*MLW*), and Mean Tide Level (*MTL*) defined as the level halfway between *MHW* and *MLW*. Mean Sea Level (*MSL*) is determined as the arithmetic mean of the hourly heights recorded during the same period. Finally, using a 19-year series of measurements that include only the highest and lowest measurements observed each tidal day, the tidal datums of Mean Higher High Water (*MHHW*) and Mean Lower Low Water (*MLLW*) are found. Again, values for all of these datums apply at only one location – the location where the tidal measurements were made that determined them.

The time-consuming 19-year average isn't everyone's choice for tidal datum definition. British hydrographers have used **Indian Spring Low Water (*ISLW*)**, a tidal datum introduced by G.H. Darwin for reckoning charted depths in India and the Middle East. This datum was defined as a downward offset from mean sea level equal to the sum of four tidal constituent amplitudes: M_2, S_2, K_1, and O_1. Mean sea level is affected by a number of factors but the 18.6-year nodal cycle is not one of them. Thus, wherever an acceptable mean sea level datum was available, the *ISLW* datum could be quickly determined from a harmonic analysis of the tide. British and other national authorities now use **Lowest Astronomical Tide (*LAT*)** in place of *ISLW*. It, too, requires an accepted *MSL* along with a more extensive harmonic analysis; *LAT* is defined in practice as the lowest astronomical tide forecast by a tidal prediction model referencing *MSL* and using a full set of harmonic constituents. In other words, where *LAT* is the datum in use, there will never be a *negative* predicted tide. Although both methods accomplish their objectives, the British approach and the U.S. approach to tidal datum definition share a common problem – changing sea level.

5.7 SEA LEVEL TREND AND THE TIDAL DATUM EPOCH

As if collection of 19-year data sets were not enough to keep government employees busy, defining secular (long-term) trends in sea level have added to their work. And if you were to say that making all these measurements at U.S. East Coast tide stations alone (some 1,642 stations from Maine to Miami) is out of the question, you'd be absolutely right. *NOAA's National Ocean Service* makes long-term tidal measurements and daily predictions at a relatively small number of *primary* tide stations in the United States (33 between Maine and Miami). Monthly mean sea level is calculated at each of these stations every calendar year and, after removing the solar annual and solar semiannual constituents[2], the results are plotted in a time series to determine the secular

---

[2] Monthly averaging effectively removes all but the seasonal tides, which are normally determined from multi-year averages of monthly mean sea level for individual months.

trend. The trend is examined to determine the need to update the current 19-year tidal datum epoch. Given the fact of *global (eustatic) sea level rise* in combination with tectonic emergence or submergence of the earth's crust at continental margins, it's a safe bet that secular trends will never be in short supply. For that reason, NOAA/NOS can be expected to continue their policy of updating the 19-year series used to compute *MSL* and other tidal datums at primary stations - updates now occurring approximately every twenty years. The series of years in use at any one time is referred to as the **National Tidal Datum Epoch**. In response to observed sea level trends, US tidal datums have been adjusted (by amounts that vary from station to station) during four National Tidal Datum Epochs: *1924-1942, 1941-1959, 1960-1978*, and most recently, *1983-2001*.

You may have noticed that I use the term **sea level trend** in place of sea level rise. Sea level rise is so much a part of our everyday language now that it may require some mental effort to realize that sea level is actually falling in a number of places: falling relative to the land, that is. A good example of falling sea level occurs at Juneau, Alaska as shown below in Fig. 5.2. Based on the monthly mean sea level data in this figure, the mean sea level trend from 1944 through 2003 at Juneau was about -12.59 mm/year (-4.13 feet/century) with a standard error of 0.22 mm/yr. Why is this happening? Juneau is located along an *active* continental margin – a zone of continuing uplift and fault-driven earthquakes.[3] Land, more than sea, is rising here.

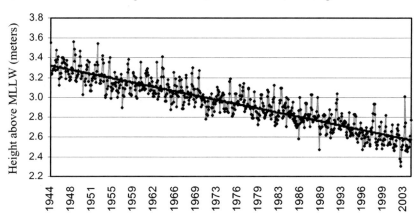

Fig. 5.2. Monthly mean sea level at Juneau, Alaska with a falling trend of –12.69 mm/year. Heights refer to 1983-2001 mean lower low water (National Ocean Service data).

In contrast to the western part of the continent, the eastern United States occupies a *passive* continental margin where there are relatively few earthquakes and coastal areas are experiencing subsidence at varying rates, aided in some locations by excessive groundwater removal. The tide station at Sewells Point in lower Chesapeake Bay furnishes an example of the *positive* sea level trend that typifies passive margins with a

---

[3] According to plate tectonic theory, this is a consequence of the North American plate overriding the Pacific plate as they converge along an active continental margin in this zone.

rise of 4.25 mm/year (1.39 feet/century) and a standard error of 0.14 mm/yr based on monthly mean sea level data from 1930 through 2003 as shown in Fig. 5.3 below.

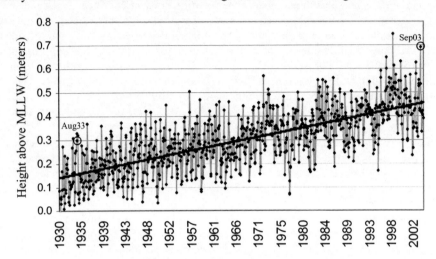

Fig. 5.3. Monthly mean sea level at Hampton Roads (Sewells Point), VA displaying a rising trend of 4.25 mm/year (solid line) from 1930 through 2003. Heights refer to 1983-2001 mean lower low water (National Ocean Service data). Two major hurricanes occurred during Aug 1933 and Sep 2003 (circled monthly means).

Sea level trends are very much a subject of debate today because of concern that global sea level rise may be accelerating as the result of **global warming**. Recent climatological evidence suggests that we may well experience a marked increase in the global rise rate within the next few decades. While it is tempting to delve into the most recent data over a decade or less in hopes of pin-pointing the moment of change, Figs. 5.2 and 5.3 should suggest to you that it is unwise to do so. The monthly and yearly extremes shown in these figures are simply too great to obtain a reliable sea level trend within that time span.

Rising sea level and storm tides - Interestingly the monthly sea level variations suggest something else. One of the chief reasons for our concern about future sea level rise is the added elevation it will impart to storm tides produced by severe winter storms and hurricanes. If we are told that the 'storm of the century' for Chesapeake Bay is one likely to recur at an average interval of 100 years with roughly the same strength each time, Fig. 5.3 tells us that, on average, the sea level from which the storm tide arises will have gone up by almost half a meter in the same interval. But that expectation is based solely on the trend line, ignoring the probability of a monthly variation in water level above or below the trend. In the worst case, we could very well see an additional 20 cm tacked on to the total for the storm tide should a hurricane chance to arrive in the wrong month. Adding it all up, the storm tide during hurricane *Isabel* in 2003 started out a good 41 cm higher in Hampton Roads than it did for another major hurricane that occurred in 1933 (circled months in Fig. 5.3). Storm tides and the storm surges that produce them are discussed at greater length in Chapter 9.

5.8 TRANSFERRING TIDAL DATUMS

Since tidal records that include all of the years specified in the National Tidal Datum Epoch are generally available only at NOS primary tide stations, how can anyone determine a tidal datum such as *MLLW* at another location that doesn't have those records? Obviously it can't be done by direct means but the equivalent of a tidal datum can be readily determined by **simultaneous comparisons**. 'Simultaneous' means that we require two separate tidal records that cover the same time period at two different tide stations. One of these is the *reference station* that already has directly determined datums, or their equivalent, and the other is the *receiving station* to which the datums are to be transferred. The choice of the time period is irrelevant but its length (duration) is not; i.e., we'll get better results from a one-year comparison than a two-week comparison. Also, the two stations have to be in fairly close proximity along connecting waterways so that they experience a similar mix of astronomical and meteorological forcing.

The basic assumption of the method of simultaneous comparisons is that, on any given day, water levels at both the reference station (Station A) and the receiving station (Station B) will undergo similar deviations from the perceived norms for the tidal epoch. For example, if the average water level calculated at Station A on a given day happens to be 20 cm above the 1983-2001 *MSL* datum, it is assumed that the average water level calculated on the same day at Station B is also 20 cm above 1983-2001 *MSL*. Of course the actual deviations may not be exactly the same on that or any other day. Statistically, we assume that the deviations at both stations have the same theoretical mean over time and that the deviations from this mean are normally distributed. If this assumption is justified, then it's just a matter of collecting a data sample of adequate length that will keep us within the error limits we're willing to accept on the transferred datum. Experience shows that the sample ideally should not include less than one month of simultaneous comparisons using appropriate sampling rates and water level averages.

Mean sea level transfer - An appropriate sampling rate for sea level computations is one sample per hour, or approximately 720 water level observations in an average month of thirty days. The arithmetic mean of these observations is the **monthly mean sea level (*MMSL*)**. If $MMSL_A$ is the monthly mean sea level for a given month at Station A, and $MMSL_B$ is the monthly mean for the same month at Station B, the assumed equality of their deviations from the *MSL* tidal datum is represented by the following equation:

$$MMSL_B - MSL_B = MMSL_A - MSL_A$$

or

$$MSL_B = (MMSL_B - MMSL_A) + MSL_A$$

where MSL_A is the accepted mean sea level datum at Station A and MSL_B is the *equivalent* datum for Station B. The key variable in the comparison is the difference in monthly mean sea level between stations. Note that the absolute elevation does not matter as long as staff zero remains fixed at both stations and is not allowed to change without adjusting the records accordingly.

Mean tide level, mean range transfer – Mean tide level (*MTL*) is defined as the level midway between the mean high water datum (*MHW*) and the mean low water datum (*MLW*) or simply the average of the two. Mean tide range (*Mn*) is defined as their difference. Although *MTL* is computationally different from *MSL*, it transfers the same way using *monthly mean tide level* (*MMTL*) in place of *MMSL*:

$$MTL_B = (MMTL_B - MMTL_A) + MTL_A \qquad (5.1)$$

where

$$MMTL_A = \tfrac{1}{2}(MMHW_A + MMLW_A)$$
$$MMTL_B = \tfrac{1}{2}(MMHW_B + MMLW_B) \qquad (5.2a,b)$$

and

$$MTL_A = \tfrac{1}{2}(MHW_A + MLW_A) \qquad (5.3)$$

The letter symbols *MMHW* and *MMLW* represent *monthly mean high water* and *monthly mean low water*, respectively, at the place designated by the station subscript. Mean tidal range (*Mn*) is transferred using the *range ratio* (*RR*) between Station A and Station B:

$$Mn_B = RR \cdot Mn_A \qquad (5.4)$$

where

$$RR = \frac{MMHW_B - MMLW_B}{MMHW_A - MMLW_A}$$

and

$$MN_A = MHW_A - MLW_A$$

MHW, MLW transfers - Equivalents of the *MHW* and *MLW* tidal datums at Station B are calculated as

$$MHW_B = MTL_B + \tfrac{1}{2}Mn_B$$
$$MLW_B = MTL_B - \tfrac{1}{2}Mn_B \qquad (5.5a,b)$$

Examples of the statistical errors associated with *MHW* and *MLW* equivalents are presented in sec. 5.9 How Good are the Results?.

MHHW, MLLW transfers - Equivalents for the *mean higher high water* (*MHHW*) and *mean lower low water* (*MLLW*) datum are obtained using essentially the same transfer methods as listed above, except that the mean tide level, mean range and range ratio all require a slight modification to reflect the singular choice of higher highs and lower lows among each day's tides. Starting at Station B, *MTL* is re-calculated as

$$MTL'_B = (MMTL'_B - MMTL'_A) + MTL'_A \qquad (5.6)$$

where

$$MMTL'_A = \tfrac{1}{2}(MMHHW_A + MMLLW_A)$$
$$MMTL'_B = \tfrac{1}{2}(MMHHW_B + MMLLW_B) \qquad (5.7\text{a,b})$$

and

$$MTL'_A = \tfrac{1}{2}(MHHW_A + MLLW_A) \qquad (5.8)$$

Next, a higher tidal range (*HMn*) and higher range ratio (*HRR*) are calculated as

$$HMn_B = HRR \cdot HMn_A \qquad (5.9)$$

where

$$HRR = \frac{MMHHW_B - MMLLW_B}{MMHHW_A - MMLLW_A}$$

and

$$HMn_A = MHHW_A - MLLW_A$$

Finally, the *MHHW* and *MLLW* equivalents at Station B are calculated as

$$MHHW_B = MTL'_B + \tfrac{1}{2} HMn_B$$
$$MLLW_B = MTL'_B - \tfrac{1}{2} HMn_B \qquad (5.10\text{a,b})$$

Higher mean range, *HMn*, has no relation to extreme high water or extreme low water as defined in tide and current glossaries and is <u>not</u> equivalent to *spring range* (*Sg*), the average range occurring at the time of spring tides. Where can you get tidal data and information about tidal datums? Online, of course. Visit http://co-ops.nos.noaa.gov where you can download verified water level observations from hundreds of NOS stations as well as a complete listing of the accepted tidal datum elevations for most of them.

As you can see from the above, the key variables in a tidal datum transfer are the *range ratio* (**RR**) and the *mean tide level difference* (**ΔMTL**) between two stations nominally using a month of simultaneous comparisons for this purpose.

5.9 HOW GOOD ARE THE RESULTS? – Aside from data quality, the answer to this question depends on the length of the series used for comparison. Suppose 'Jane', a landowner living on the York River near Gloucester Point, Virginia, is interested in locating the mean high water mark on her property. Jane has read about the method of simultaneous comparisons and has installed a tide staff on a nearby pier. 'Joe' is Jane's next-door neighbor and he is sharing Jane's tide staff to find *MHW* on his own property. Neither can afford a tide gauge but Joe, always the keen observer, is willing to read and record high and low water heights from the staff, day and night, for one month. Jane, however, wants a quick estimate of *MHW* and plans to make observations over a single day that includes two highs and two lows. How will her results compare with Joe's?

We will answer this question through statistical inference using high and low water heights from an NOS tide station at Gloucester Point in a year long comparison with simultaneous highs and lows from the NOS primary tide station at Sewells Point in Hampton Roads[4].

Table 5.1 Simultaneous Comparisons

2002	ΔMTL(m)	RR
Jan	-0.844	0.956
Feb	-0.850	0.952
Mar	-0.836	0.936
Apr	-0.832	0.960
May	-0.839	0.952
Jun	-0.832	0.969
Jul	-0.830	0.969
Aug	-0.832	0.950
Sep	-0.841	0.953
Oct	-0.846	0.939
Nov	-0.841	0.938
Dec	-0.847	0.959
mean	-0.839	0.953
stdev	0.0066	0.0109

As previously explained, just two pieces of information are needed to transfer the *MHW* datum from Sewells Point (Station A) to Gloucester Point (Station B). These are the *mean tide level difference*, ΔMTL, and the *tidal range ratio*, *RR*. The table at right contains twelve monthly averages for these two variables, plus their annual mean and standard deviation for 2002. The standard deviations in the bottom row provide a handle on the inferred error for ΔMTL and *RR* drawn from this sample; i.e., if Joe had picked one of the twelve months at random, he would stand a roughly one-in-three chance of missing ΔMTL, as estimated by the 2002 annual mean, by 0.66 cm or more in either direction. He would have the same chance of missing the *Mn* ratio by approximately one percent or more in either direction[5].

What about Jane's single day results? Without showing all the daily averages for 2002 in a table like the one above, the annual means remain the same but the standard deviations for ΔMTL and *RR* increase to 0.0267 and 0.0342, respectively. Thus Jane's random pick has approximately a one-in-three chance of missing ΔMTL by 2.7 cm or more and the same chance of over (under) estimating *RR* by 3.4 percent or more.

---

[4] Heights at both stations refer to station datum; i.e., staff zero.
[5] Given a normal distribution in which 68% of all samples fall within one standard deviation of the mean.

MHW error estimates – Table 5.1 provides all the information we need to determine the quality of the comparison in terms of monthly and daily error estimates for *MHW*, the datum both Joe and Jane hope to find on their property at Gloucester Point. An estimate of the transferred datum is calculated using Eq. (5.5a). If we assume that the MTL_A and Mn_A values at Sewells Point contain no error, then from Eqs. (5.1) and (5.4), it is clear that the errors associated with MTL_B and Mn_B are equivalent to the errors for ΔMTL and the range ratio, RR, which we already have. To find the error that propagates from these variables to MHW_B through Eq.(5.5a), we combine variance (square of the standard deviation) in the following way:

$$\text{var}\{MHW_B\} = \text{var}\{\Delta MTL\} + \tfrac{1}{4}\text{var}\{RR\}$$

Thus, after measuring highs and lows for <u>one month</u>:

$$\text{var}(MHW_B) = (0.0066)^2 + \tfrac{1}{4}(0.0109)^2$$
$$\text{stdev}\{MHW_B\} = 0.0086\ m = 0.86\ cm$$

and after measuring highs and lows for <u>one day</u>:

$$\text{var}(MHW_B) = (0.0267)^2 + \tfrac{1}{4}(0.0342)^2$$
$$\text{stdev}\{MHW_B\} = 0.0317\ m = 3.17\ cm$$

These standard deviations represent *precision estimates* for both MHW_B and MLW_B, the datum elevations found on Jane's staff at Gloucester Point by combining the annual ΔMTL and RR values for 2002 with $MTL_A = 1.637$ m and $Mn_A = 0.753$ m, the accepted NOS mean tide level and tidal range values at Sewells Point; i.e.,

$$MTL_B = \Delta MTL + MTL_A = -0.839 + 1.637 = 0.798\ m$$
$$Mn_B = RR \cdot Mn_A = (0.953) \cdot (0.753) = 0.718\ m$$
$$MHW_B = MTL_B + \tfrac{1}{2}Mn_B = 0.798 + \tfrac{1}{2}(0.718) = 1.157\ m$$
$$MLW_B = MTL_B - \tfrac{1}{2}Mn_B = 0.798 - \tfrac{1}{2}(0.718) = 0.439\ m$$

Now that our 2002 datum transfer is complete, we'll compare the results with the corresponding datum values that NOS has stored in its database for Gloucester Point. Those results are presented for comparison with the numbers we have just calculated in Table 5.2 below.

Table 5.2 Comparison of transferred datums with NOS datums.

Datum (m)	2002	NOS	difference
MHW	1.157	1.173	-0.02
MTL	0.798	0.808	-0.01
MLW	0.439	0.439	0.00
Mn	0.718	0.734	-0.02

Since NOS has the official say on datum values at all U.S. tide stations, the difference column in the above table may be regarded as a practical test of the *accuracy* of our 2002 simultaneous comparison. Because we slightly underestimated both *MTL* and *Mn*, our comparison tends to favor a *MLW* agreement at the expense of a better match with *MHW*. But, as Joe's example shows, the precision of a one-month comparison (± 0.9 cm) for either datum is certainly on a par with accuracy results for 2002. In fact, considering Jane's precision for a one-day comparison (± 3.2 cm), her own *MHW* determination should offer a fairly good first approximation - and an even better one might result if she could just be persuaded to keep on observing highs and lows for another day or two!

6

Tides and currents in a large estuary

Tides reach the coastal zone, including its bays and estuaries, from the open ocean. As explained in Chapter 3, the North Atlantic ocean basin is home to a large rotary (Kelvin) wave generated by global scale, tide-producing forces. However, the boundary that confines this wave is not the coastline but the continental shelf break where water depths abruptly deepen - from 100 meters to more than 3,000 meters off the central US East Coast. As the tide wave crosses this boundary from east to west (see Fig. 3.12), it enters what is in effect a new and much shallower basin on the continental shelf. In a cross-section drawn normal to the coastline, the basin appears closed at one end (the shoreline) and open at the other (the shelf break) with length equal to the shelf width. In Chapter 3 we learned that an open basin of this kind responds to incoming waves with a unique water level oscillation based on its dimensions (width and depth) and the period of the semidiurnal tide. Because it has virtually no tide of its own but simply responds to constant outside forcing, the shelf 'basin' is referred to as a **co-oscillating tidal system**. What does this mean? In a nutshell, it means that while the tide range at the shelf break remains largely the same, the tide range along the coast will vary widely in response to changes in shelf width and depth offshore. We can expand on that concept to explain in simple terms why tides and currents in adjoining bays and estuaries behave as they do.

Co-oscillating tidal systems are by no means restricted to the continental shelf. Where the coastline opens into a bay, the bay likewise responds to the external tide, as does its tributaries in hierarchical fashion as the dimensions of the connecting systems decline. Let us start our trek at the entrance to the largest estuary in the United States: Chesapeake Bay.

6.1 TIDES INSIDE CHESAPEAKE BAY

The tide's movement into Chesapeake Bay and its tributaries (*river estuaries*) bears some similarity in origin to a wind wave traveling away from an ocean storm center toward some other area that it will eventually reach. Indeed, the tide entering the Bay acts like the long progressive wave it is with its crest marking a high tide and its trough marking a low tide at later and later times as the wave sweeps past points farther and farther into the Bay. But tide waves do not de-couple themselves from the engine that creates them, as wind waves do when they travel outside their generating area as free waves. Chesapeake Bay itself is too small to be a generating area for tides but as long as it remains open, no tidal oscillation inside it will ever be free from the tidal 'hammer' perpetually at work at its entrance. Nothing will change its beat (tidal period) or magnitude (tidal range) there regardless of what takes place inside the Bay. Like two pendulums of similar period, one driven by the other in synchronized motion, the Atlantic Ocean shelf waters and Chesapeake Bay waters represent co-oscillating tidal systems.

6.2 CO-RANGE and CO-TIDAL CHARTS

The basic features of the Chesapeake Bay tidal system can be summarized using two charts, one containing **co-range lines** (Fig. 6.1) and the other containing **co-tidal lines** (Fig. 6.2). Tidal range is constant along a co-range line and tidal phase is constant along a co-tidal line. Looking at the first of these maps, we see a tidal range of 90 cm (3 feet) at the entrance gradually decreasing to a minimum of less than 30 cm (1 foot) some two-thirds of the distance up the Bay before increasing again to slightly more than 50 cm at the head. Of the four major tributaries, one is relatively short (York River) and displays a uniform increase in tidal range upstream while the longer ones (James, Rappahannock and Potomac Rivers) show a slight decrease in range before the final increase upstream.

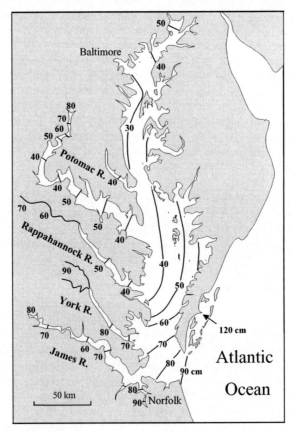

Fig. 6.1. Co-range lines (cm) for observed tides in Chesapeake Bay (After Browne and Fisher, 1988). A tidal range of 120 cm occurs behind Virginia's barrier islands.

The tidal range at the head of all four tributaries approaches the range at the bay entrance itself. Outside the Chesapeake Bay a maximum tidal range of 120 cm (4 feet) is found inside the shallow lagoons and marshes behind Virginia's barrier islands.

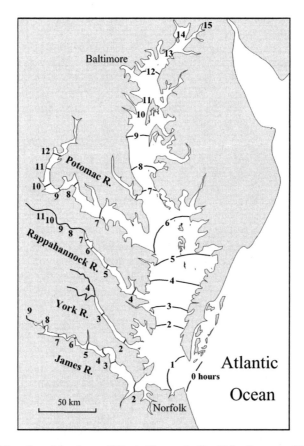

Fig. 6.2. Co-tidal lines (hours) for observed tides in Chesapeake Bay (After Browne and Fisher, 1988).

The co-tidal lines shown in Fig. 6.2 depict the *phase* of the tide, in hours, relative to its phase at the Chesapeake Bay entrance. In viewing this map, we should imagine a tide wave entering the Bay with high tide occurring at all points along its crest. At hour zero, the crest position is defined by the '0' line at the entrance whereas one hour later it is positioned along the line marked '1' and so on as the wave moves up the Bay. We can see that, by hour 12, the crest of one wave will have traveled beyond Baltimore, just as the crest of the next wave is about to enter the Bay (on average, a new crest arrives there every 12.42 hours). The Chesapeake Bay is one of the few estuaries in the world large enough to contain a complete tide wave at all times. But as the bay is open at one end and closed at the other, wave reflection occurs; depth, width, and bottom friction, as well as the earth's rotation, cause the tide wave to change in important ways.

6.3 A SIMPLE SHALLOW-WATER WAVE MODEL
Before presenting this model which attempts to describe some of the interesting things that happen to tide waves on their way to the head of the Bay (and back down the Bay after reflection at the head) you and I should be aware of a not so small caveat. It's right

about here in an advanced oceanography course that the instructor brings out the equations of motion and begins to talk about whether a term is important or not in a given situation and whether we must keep it or can throw it out. Since these equations have quite a few terms, everyone is glad to learn that this one or that one can be ignored. Glad because the 'simplified' equations become easier to solve so that we get answers that tell us precisely how the system works at a given place and time. But that's another level of inquiry. We can get a very useful *basic* understanding of how things work with just one very simple equation – the *shallow water wavelength equation* introduced in Chapter 3:

$$L_s = T\sqrt{gh} \qquad (6.1)$$

As you recall, eq. (6.1) relates the wavelength in shallow water, L_s, to the water depth, h, and the wave period, T. The remaining symbol, g, represents the acceleration of the earth's gravity introduced in Chapter 2. It's this equation that tells us, once we know the depth and the period of the tidal constituent we're using, what size wave we have and whether it will fit within Chesapeake Bay or some other body of water that's closed at one end and open at the other. Ready to see where this leads? Good. But first we'll deal in a qualitative way with the Coriolis term – one of the ones that can't be chucked out of the equations of motion for a body as large as Chesapeake Bay.

Coriolis effect on tidal range – As it does in all large water bodies, the tide wave inside Chesapeake Bay responds to the *Coriolis effect*. The *longitudinal* decrease in tidal range proceeding up the Bay is accompanied by a *transverse* increase in range across the Bay (Fig. 6.1) - from a minimum near the western shore to a maximum near the eastern shore at any given latitude - due to the Coriolis effect caused by the earth's rotation. Neglecting other effects, a major current (one involving a lot of water in motion) tends to turn to the right in the northern hemisphere. This means that a northward moving current inside Chesapeake Bay will cause water to accelerate and move towards the eastern shore. A maximum northward current occurs as the crest of the incoming tide wave passes a given latitude (see Fig. 3.5). The Coriolis effect will then cause water levels to rise slightly higher against the eastern shore than they would otherwise during high tide. A little more than six hours later, the wave trough will pass the same latitude producing low water and a maximum ebb current flowing to the south. Coriolis force acting on the southerly current will tend to move water to the west causing water levels along the eastern shore to fall below the normal low that would occur in the absence of the Coriolis effect. The net result, higher highs and lower lows, is an increase in tidal range along the eastern shore. Along the western shore the opposite occurs (lower highs and higher lows) and the tidal range there is reduced.

Wave attenuation and reflection – Although the notion of a progressive wave agrees well with the pattern of co-tidal lines shown moving up the Chesapeake Bay and its tributaries in Fig. 6.1, we should not get too comfortable with this concept alone since waves have at least two other properties – *attenuation and reflection* - that work against it. Waves traveling over long distances in very shallow water will dissipate much of their energy through bottom friction and undergo attenuation, a drop in amplitude. What do we mean by 'very shallow'? This is an added nuance to the standard definition of a shallow water wave; i.e., a wave whose length is long in comparison to the water depth. All tide waves are shallow water waves in that sense. But when the depth is so small that it begins to compare to the tide wave amplitude, that's very shallow and attenuation becomes important. That process agrees well with the initial decrease in tidal range shown in Fig. 6.1, but not so well with the subsequent increase in range that appears near the ends of the waterways. That brings up the other process – *wave reflection*. If the tide wave is not fully attenuated by the time it reaches the end of a waterway, as is clearly the case in Chesapeake Bay and its tributaries, then wave reflection is likely to occur at the landward boundary, sending what's left of the wave back in the direction it came from.

The two wave processes acting together are conceptualized in Fig. 6.3 with the M_2 tide wave ($T = 12.42$ hours), assuming a constant depth ($h = 5$ m) and an exponential amplitude attenuation of 90% from bay mouth to bay head and back. From eq. 6.1 we have $L_s = 313$ km. The light dotted line shows how the incoming wave would have

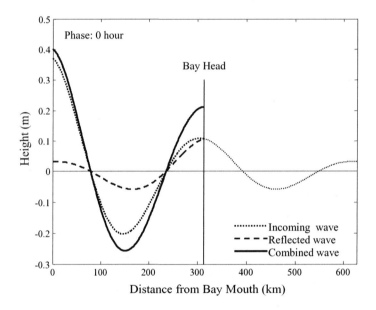

Fig. 6.3. Diagram showing combination of incoming and reflected M_2 tide wave with wave attenuation in effect (5 m depth).

looked had there been no bay head to reflect it back toward the ocean. The solid line, the combination of the incoming (dotted) and reflected (dashed) wave, is the one of interest because it defines the actual shape of the water surface profile up and down the Bay at a particular moment in time, phase hour '0' in this case. We know that an hour later, at phase hour '1', the shape will be different since the incoming and reflected waves that combine to produce it are constantly in motion.

To see the net effect for M_2, we could calculate the shape of the combined wave (the solid line in Fig. 6.3) and plot its position every hour over a 25-hour period (about one lunar day) on the same graph. But to do things one step at a time, suppose we first consider what the 25 superimposed waveforms would look like assuming Chesapeake Bay to be frictionless with no wave attenuation along its entire length. This scenario is imaginary of course, but we should not be too surprised at what appears in Fig. 6.4a: namely a standing wave system with two nodes and three antinodes (one antinode at each end and one antinode in the middle of the Bay).

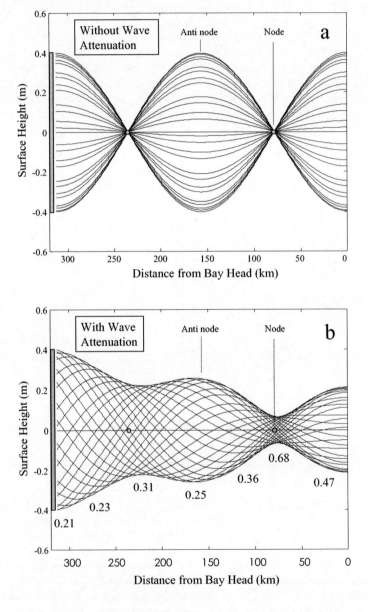

Fig. 6.4. Combined M_2 tide wave plotted hourly over a lunar day without attenuation (upper plot a), and with attenuation (lower plot b). Tidal form numbers appear below the tidal range envelope of the lower graph.

The next step is the crucial one. The mesh-like structure in Fig. 6.4b shows what happens after adding wave attenuation to the standing waves depicted in Fig. 6.4a. Now the envelope containing the waves shows maximum tidal range at only one location (the bay mouth) and minimum tidal range at only one location (a sector about 75 km from the bay head). Tidal form numbers appearing below the waveforms in Fig. 6.4b clearly indicate an inverse relationship with the M_2 tidal range, revealing that the greatest tendency for a diurnal tide occurs near Baltimore (form number 0.68) where tides are *mixed, mainly semidiurnal* in contrast to Norfolk (form number 0.21) where tides are *fully semidiurnal*. Helping to shape these numbers is the fact that L_s wavelengths for the diurnal tidal constituents K_1 and O_1 are roughly twice that of M_2 (their average period is twice the M_2 period). Had these constituents been shown in Fig. 6.5, they would have depicted a standing wave with only one node near the middle of the Bay and a relatively large antinode near the head. This is consistent with the observation that the largest form numbers are found between Annapolis and the head of the Bay, recalling that the tidal form number is computed as the ratio of the diurnal constituent amplitudes ($K_1 + O_1$) to the semidiurnal constituent amplitudes ($M_2 + S_2$).

While the form number serves as a useful guide to the character of the tide in Chesapeake Bay, so does the mesh pattern seen in Fig. 6.4b. Notice that the mesh is *diamond-shaped* until you get to within about 50 km of the bay head where the lines representing water level straighten out and start to become parallel. The diamond shape suggests horizontal movement characteristic of progressive waves, a pattern wholly consistent with the co-tidal lines in Fig. 6.2 that advance northward through the lower and middle bay regions. This also means that the bay entrance is no longer an antinode and strong currents can exist there. The parallel lines on the right in Fig. 6.4b imply standing wave conditions in the uppermost sector of the Bay. That can account for some of the increase in tidal range depicted by the co-range lines in Fig. 6.1 but the continued presence of co-tidal lines suggests that a progressive wave element persists all the way to the end.

By the way, did you notice that the distance scale is reversed in Figs. 6.4a and 6.4b? I do that to remind myself that the *head* of an embayment, not the mouth, is the proper starting point for measurements aimed at fitting a tide wave within a semi-enclosed water body and working out a rough approximation of the tidal dynamics inside it. For example, suppose Chesapeake Bay happened to be only 230 km in length as opposed to 313 km. If that were the case, then the result should look something like Fig. 6.5 as shown below. The key to understanding this figure rests with the fact that the closed end of the embayment must always coincide with an antinode while the mouth must have the same tidal range as the adjacent ocean[1].

Bound by the above two constraints and assuming the effective depth is still the same (5 m), the *shape* of the tidal range envelope in Fig. 6.5 remains unchanged but with part of one end (the seaward end) cut off. However, there is one major difference: the range at every point inside the bay has doubled for this scenario compared to the original in Fig. 6.4b! This change is evident noting the doubling of the vertical scale on the left side of the graph in Fig. 6.5, observing that the M_2 tidal range at the bay mouth is still approximately 0.8 m but now adjoins the envelope at the 230 km distance as indicated

---

[1] Or approximately so, allowing for some adjustments in the vicinity of the entrance.

in the figure. Clearly the tidal range inside a waterway is sensitive to the waterway length as well as depth.

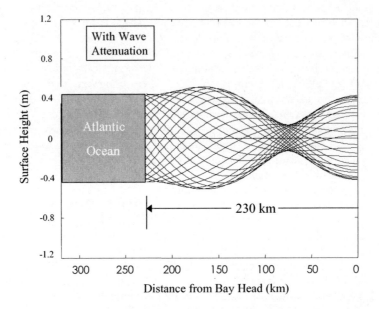

Fig. 6.5. M_2 tide wave plotted hourly over a lunar day in a 'shortened' Chesapeake Bay (5 m depth).

Getting back to the aforementioned caveat, I was once part of a group working to develop a hydrodynamic model of the York River in Virginia. Our efforts to calibrate the model so that its predictions matched the observed tidal range at certain key locations were not successful at first, even after employing all the right equations and making the usual adjustments (bottom friction coefficients, grid cell dimensions and associated depths). In fact, nothing worked until it was realized that the representation of the two tributaries to the upper York, the Pamunkey and Mattaponi Rivers, were not quite as long as they should have been in the model domain. It is the nature of these rivers to continue inland as narrow tidal streams with no clearly defined endpoint, making a final determination of their length by inspecting a map somewhat arbitrary. However, once new endpoints were chosen giving additional lengths to both rivers, predicted tidal ranges quickly fell into line with observed ones. However at that point the numerical model could give results that our simple wave model could not; for example, correct prediction of the added increase in tidal range observed at the heads of the major river tributaries (Fig. 6.1). The reason is that *progressive shoaling* and *channel narrowing* cause an additional increase in range to occur in regions where progressive wave behavior is still present as evidenced by regularly advancing co-tidal lines (Fig. 6.2).

The Elizabeth River, a short waterway – Does the tide behave differently in any of the shorter tributaries within Chesapeake Bay? It does if the depth is relative large compared to its length. An example of just this kind of waterway is the Elizabeth River, a branching tributary to the James River that extends between the cities of Norfolk, Portsmouth, and Chesapeake, Virginia (Fig. 6.6). This river is, is effect, a tidal basin into which fresh water runoff is added from four major branches (including the Lafayette River) and numerous small streams. During a recent model study of the Elizabeth River, simultaneous measurements from seven tide stations (identified by four-letter station codes in Fig. 6.6) were analyzed that quickly confirmed the existence of a *standing wave*.

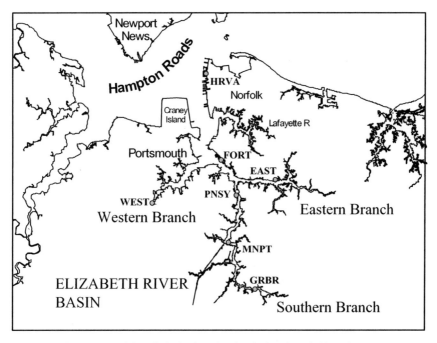

Fig. 6.6. Map of the Elizabeth River showing the location of tide stations.

To find out just what was 'confirmed', a quick look at the system's basic dimensions is helpful. The main stem of the Elizabeth River runs from its entrance at Sewells Point in Hampton Roads, Virginia (tide station HRVA) to the end of the Southern Branch near the waterway locks at Great Bridge (tide station GRBR). This segment is about 30 km long and, including both shoal margins and a deep navigation channel, has an average depth of about 4.3 m. Its length is thus quite short compared to the corresponding shallow water wavelength (L_s=290 km). In fact, it is slightly less than half the one-quarter **critical resonance length** ($0.25L_s$=72 km) for a waterway of this depth[2]. Again measuring from the *end* of the waterway, not the mouth, the tidal range variation inside the Elizabeth River should be as shown in Fig. 6.7.

---

[2] Theoretically the range inside a waterway containing one-quarter of a standing wave would be infinite in the absence of friction.

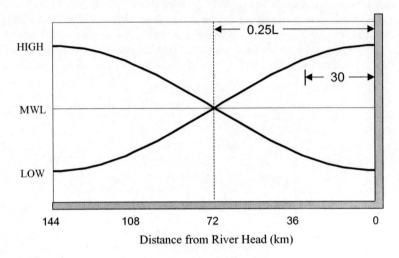

Fig. 6.7. Standing wave in a channel 4.3m deep (T=12.42 hours).

As the above figure shows, the expected tidal motion would be that of a standing wave with a period of 12.42 hours, or more correctly, the last 30 km of that wave. The tidal range produced by such a wave would steadily increase over the 30 km distance from the mouth of the Elizabeth River to its head. And because it's a standing wave we're talking about, the times when the water level extremes (high tide or low tide) occur should be nearly the same throughout the waterway. The question is, do the measurements confirm this picture?

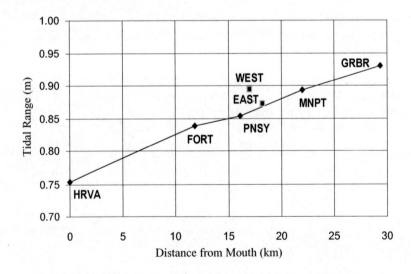

Fig. 6.8. Mean Tidal Range (1960-1978 Tidal Epoch), Elizabeth River.

They do. Fig. 6.8 contains a plot of the measured tidal range values obtained from the seven tide stations available in the Elizabeth River. As expected, the ranges show a steady increase from about 0.75 m at station HRVA to 0.93 m at station GRBR. Records from a pair of tide stations in the Eastern Branch (EAST) and Western Branch (WEST) of the Elizabeth River confirm an increase in range moving away from the main stem toward their respective heads. As for the phase, the records from the seven stations indicate that, during an average tide, high waters throughout the Elizabeth River system occur within less than 25 minutes of one another; low waters occur within 15 minutes of one another or less.

The tidal ranges in Fig. 6.8 were calculated using *simultaneous comparisons* as described in Chapter 5. Data collected for a common period of time (approximately three months) were compared between each station shown in Fig. 6.6 and Hampton Roads (station HRVA), a NOAA *primary* tide station reporting the accepted tidal range of 0.753 m for the then current Tidal Epoch (1960-1978). Initially a *range ratio* was calculated between each 'secondary' station and the primary station at HRVA using the simultaneous data. Then the accepted range at station HRVA was multiplied by the range ratio for each secondary station. The resulting 'transfer' of the accepted tidal range to each of the other stations produced the equivalent of the 1960-1978 range, even though they had no data at all in the defining time period. However, had we simply used the short-term average of the tidal range for each station alone, the results would be quite misleading because they would ignore the long-term cycles discussed in Chapter 2. The method of simultaneous comparisons is thus a very useful technique for transferring tidal range in addition to tidal datum elevations as noted in Chapter 5.

6.4 TIDAL HEIGHT AND TIDAL CURRENT RELATIONSHIPS

Knowing in advance what the tides and tidal currents are going to do is essential if you own a cruising boat and plan an outing on one of Chesapeake Bay's tidal waterways. Tidal height predictions for a given time of day can decide whether you have enough water to pass through critical areas – areas such as the 'Swash Channel', a narrow shortcut between the Lower York River and Mobjack Bay. Predictions of the direction and strength of tidal currents can help you estimate how much time will be needed to get to your destination and back. Although there's no substitute for having both tide and current predictions aboard as you get underway, it doesn't hurt to familiarize yourself with the relationship between the two, especially in a body of water that you visit often. For example, if you witness high water at such and such a point, can you infer what the current might be doing there? You can if you know the general relationship and that comes down to whether the tide is behaving like a progressive wave, a standing wave, or something in between the two.

To find out, all you have to do is compare a predicted tide curve with a predicted current curve. That's an easier job than you might think thanks to NOS tide and current prediction tables that contain this information for quite a few Chesapeake Bay locations. In recent years, printed copies of the tables themselves have become harder to find, but the information they contain is available on the NOAA/NOS Web Site (see Chapter 8 for a description of the data available and where to look for it). Even better for those who have a personal computer or laptop, nautical software packages are now available at boating stores that actually draw the tide and current curves for you. They're considered very reliable as far as the predictions are concerned because they use NOS-determined values for the tidal constituents described in Chapter 4.

Predictions, in fact, confirm the concept previously introduced - that the tide wave entering Chesapeake Bay propagates through its lower main stem predominantly as a progressive wave, but one that starts to turn into a standing wave as it reaches the bay's major tributaries. Four NOS stations will be used here to illustrate the process graphically, one at Fishermans Island on the north side of the Chesapeake Bay entrance and three covering the York River tributary at the positions shown in Fig. 6.9. Fig. 6.10 shows the predicted tide and current curves for theses four stations during a full moon on April 26, 2002.

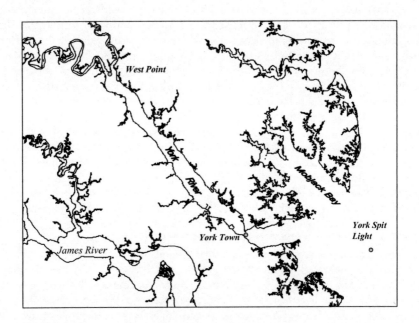

Fig. 6.9. Map of the York River showing three NOS stations used to illustrate relationships between tide and current predictions in a river estuary.

The two curves for the first station at Fishermans Island (uppermost panel, Fig. 6.10) should, of course, be in phase and coincide with one another if the tide wave there is fully progressive (Chapter 3, Sec. 3.4). This is essentially what we have except that the times of maximum flood current occur slightly ahead of the times of the high waters on the tide curve. The fact that the maximum ebb current does not have this lead and effectively coincides with low water at Fisherman Island means that the current curve is *asymmetrical*, an indication that *overtides* (Chapter 4, Sec. 4.5) are present. Overtides as a rule are more pronounced in currents than in water levels. Because of the way the asymmetry presents itself in the current extremes, it is better to compare the times of high water with the times of current reversal; i.e., times of *slack water*. If we follow this convention and compare the time difference between high water and slack before ebb in Fig. 6.10, we note that York Spit Light just outside the York River entrance has essentially the same difference as Fisherman Island. But as we enter the York River moving toward West Point, the difference drops and begins to approach zero and thus standing wave conditions.

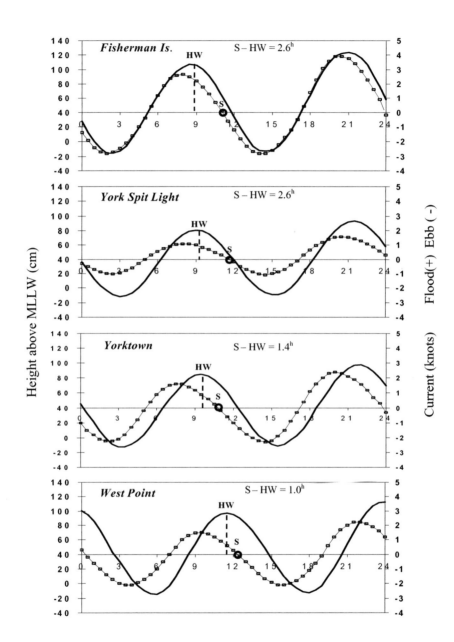

Fig. 6.10. Simultaneous predictions of tide (solid line) and current (dotted line with symbols) at four NOS stations in Chesapeake Bay on April 26, 2002. A standing tide wave is present when high water (HW) coincides with minimum current or slack (S).

6.5 ROTARY CURRENTS IN LOWER CHESAPEAKE BAY

Within a narrow waterway such as the York River, tidal currents are constrained by channel configurations and the shoreline. *Reversing currents* appear in these systems because water flows in just one direction during one-half of the tidal cycle and then 'reverses' to flow in the opposite direction during the other half. By convention, we call the landward or upstream movement *flood* and the seaward or downstream movement *ebb*. Reversing currents are the norm in tidal tributaries and for that reason we are more familiar with them on a day-to-day basis. An experienced boater leaving Sarah Creek, a tributary of the York River, will cast an eye at the base of the green day marker on his right and note whether a wake appears on the landward, seaward, or neither (slack water) side of the piling to determine the phase of the tidal current. But what happens out in the open Bay - away from shorelines and narrow channels that restrict the flow direction - is a different story. Conditions there often favor a **rotary current** – a current that flows continually while the flow direction progressively rotates through a full circle during each tidal cycle.

Although I had studied the principles behind rotary currents at sea with the US Coast & Geodetic Survey and later at the Virginia Institute of Marine Science (VIMS), I was still surprised by my first set of measurements showing what they looked like in lower Chesapeake Bay. I obtained them with an instrument tripod placed on the bottom at the seaward edge of a broad shoal. The shoal, known as the *Horseshoe*, extends three nautical miles east of Buckroe Beach in Hampton, Virginia. A small electromagnetic sensor was mounted under the tripod that recorded horizontal current speed and direction at a point about a meter above the seabed. It did this for weeks all on its own; I only had to go out once a month to retrieve the data. The *polar plot* on the right in Fig. 6.11 shows the results of the first 12 hours of measurement at the *Horseshoe*.

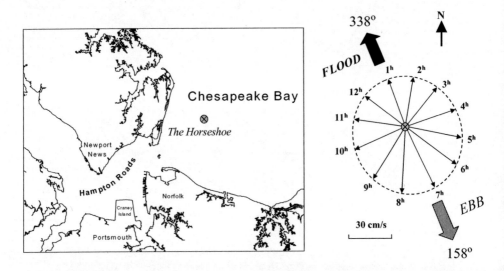

Fig. 6.11. Rotary bottom currents observed at the *Horseshoe* in lower Chesapeake Bay, September 27, 1988. Current speed and direction at each hour is indicated by the length and direction of the arrows appearing inside the dotted line. Large arrows show 'principal axis' directions for flood and ebb.

Two things about the 12 hourly current *vectors* shown in Fig. 6.11 are unusual. First of all, no vector represents a current speed less than 30 cm/s (0.6 knot) or more than 46 cm/s (0.9 knots). In fact, if we adjusted the vectors by subtracting the *average* (non-tidal) current - a flow of about 8 cm/s that occurred in the ebb direction that day - all of them would have just about the same length. The nearly circular shape of the **current ellipse** (dotted line) that encloses them underscores this point. Thus the speed of the near-bottom tidal current at *Horseshoe* shows very little change during a tidal cycle and certainly never comes to a halt so that the words *'slack water'* could apply. This is the reason why NOS current tables frequently use the words *'minimum before flood'* and *'minimum before ebb'* in place of the term 'slack water'. But near the sea floor at the *Horseshoe*, even the minimum is hard to find!

The second thing that is surprising at first is that the change in current direction in Fig. 6.11 is *clockwise*. As shown in Chapter 3, the current associated with a rotary wave in the northern hemisphere changes direction in a counter-clockwise pattern (see Fig. 3.10). Rotary currents in the open ocean follow this pattern but currents associated with progressive tide waves in an elongated embayment with shallow margins and deep central channel can go either way. According to one authority, this mix of depths causes the celerity of the waves to vary in the transverse direction, an action that leads to transverse, east-west compensatory currents being set up in addition to the north-south current of the progressive wave. Omitting the details of the theory, it predicts that clockwise rotary currents will be mostly found near the western shore while counter-clockwise rotary currents will occur near the eastern shore of lower Chesapeake Bay. Bottom currents measured at the *Horseshoe* agree at least with the first part of this prediction.

As we might expect, the *shape* of the rotary current ellipse takes on more of an elongate or 'stretched' form as the tributaries are approached. As this transition occurs, the long axis of the ellipse will generally coincide with the flood and ebb directions noted in the NOS current tables. This poses no difficulty well inside the tributaries where the shape of the current ellipse is long and narrow. However, where the shape of the current ellipse is more circular, as it is in Fig. 6.11, determination of the flood and ebb directions becomes more difficult. After taking away the average (non-tidal) current, one direction is almost as good as another for defining flood and ebb at the *Horseshoe*.

Because NOS current tables are designed with vessel navigation in mind, their daily predictions emphasize the strength of the maximum current in separately averaged flood and ebb directions. These directions do not, as a rule, differ by $180°$ as would be the case for currents aligned with a single axis. For scientific purposes, the **principal axis** of the current ellipse – a straight line with reciprocal headings - is a better tool because it provides a simple connection between diagrams that use sinusoidal waves to represent currents (e.g., Fig. 6.10) and those that use vectors for this purpose (e.g., Fig. 6.11). The principal axis aligns itself with the greatest variation in current speed among all possible directions. There is a precise mathematical technique called **eigenvector analysis** (see Chapter 10, Sec. 10.6) for finding the principal axis but at least two weeks of current measurements are needed to do it accurately. At the *Horseshoe*, a principal axis with reciprocal headings of $158°$ and $338°$ was found (Fig. 6.11).

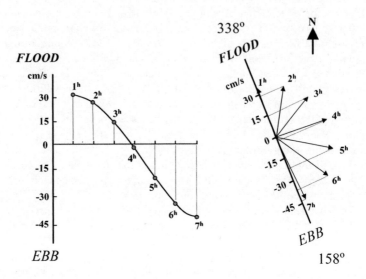

Fig. 6.12. Rotary bottom currents observed at VIMS station *Horseshoe* in lower Chesapeake Bay, September 27, 1988. Current speed and direction for hours 1-7 are projected onto the principal current axis (right) to obtain a 'reversing' tidal current curve (left).

To obtain a reversing tidal current curve, all that is needed is to project each vector onto the principal axis and note whether the resulting current component is flood or ebb-directed as shown in Fig. 6.12. The components are then used to draw reversing, sinusoidal current curves like those in Fig. 6.10. While such curves are convenient to use, you still have to remember that 'zero' readings at the points where the current component changes from flood to ebb and back again along the principal axis do not necessarily indicate slack water in all other directions.

What type of work would require the additional detail, the extra information found in a rotary current diagram? Tidal transport studies frequently do. Marine biologists investigating seasonal migrations of larval and post-larval blue crabs know that these slow swimming organisms take advantage of tidal currents to get where they need to go. It's also known that they engage in vertical migration – swimming from surface to bottom and back again – to reach currents that favor transport in different directions at different times, much like a commuter catching an 'uptown' or a 'downtown' bus depending on the time of day. And, of course, the current ellipse at the bottom may differ considerably from the one at the surface so that the 'bus schedules' at the two levels differ as well. Whether a crab knows flood from ebb – the way humans define these terms at least - is uncertain but scientists studying its movement in both time and three-dimensional space are beginning to understand the details of their travel using *hydrodynamic modeling* to obtain a more complete history of current direction and speed in their estuarine environment.

6.6 NON-TIDAL CURRENTS
Added to the tidal currents caused by the lunar and solar tractive forces are the so-called *non-tidal currents* that arise from other forces acting on the water. High winds

are responsible for *transient currents* that occur briefly and more or less at random in the Bay and inner shelf region. Although these flows are quite strong at times, they tend to have few predictable attributes going forward in time. More organized currents occur as a result of fresh water flowing into Chesapeake Bay through its numerous tributaries or *river estuaries,* as we might call them. Here we find a different type of flow called a **density current** or **gravitational flow**. This type of flow is the engine for estuarine circulation. While density currents are present on a regular basis in estuaries, their spatial and temporal features are less predictable than those of the astronomical tides and currents. But before characterizing them further, we need to understand how they arise.

Differences in density between salt water from the ocean and river water from the head of an estuary provides the driving force needed to sustain a density current. The force derives from a *pressure gradient*; the same kind that causes water to flow downhill, only here the gradient is created by differences in water density, not height. Fresh water driven in this way tends to flow seaward in a layer spread above the salt water. As mixing occurs between the fresh water layer above and the salt water layer beneath, some of the salt water becomes entrained in the seaward flow and adds to the total volume of water passing out of the estuary (fresh water plus entrained salt water). Setting aside the waters brought in and out by the tide or the wind, the *net* result is a continual loss of water volume for the estuary as a whole. Consequently, as mixing continues, bottom water must continually move landward, bringing in more salt water to make up the deficit in water volume (mass). Oceanographers refer to this as the *conservation of mass principle*.

What controls a density current? What speeds it up or slows it down? Obviously the river inflow rate is a key factor but tides also play a role. When river inflow is high, the degree of *stratification* (density difference between layers) becomes large, as does the potential for density-driven flow. More energy is then required to accomplish mixing, a breaking down of stratification that can be either partial or complete. This energy is provided by the tide through tidal current gradients that induce shear and 'tidal stirring' between layers. The ratio of tidal flow to river inflow provides a convenient index to discriminate between estuaries with different degrees of mixing. Because it falls in the mid-range of that index, the James River in Chesapeake Bay is classified as a *partially stratified* river estuary.

How do we separate estuarine currents – currents induced by tides, winds and water density differences all occurring at once? Basically it's a process of elimination. As explained in Chapter 4, the representation of tides and tidal currents requires nothing more than a collection of *cosine* waves added together to produce the so-called astronomical tide or current. Averaging a wave over its period is one way of eliminating it. If we average the tide readings in Fig. 6.10 over a complete tidal cycle, the mean water level results for that cycle; if we average the current readings in the same figure, the mean non-tidal current results. The trick is to find a cycle that all the tidal constituents involved have in common, recognizing that the curves in Fig. 6.11 include both *semidiurnal* (twice-daily) and *diurnal* (once-daily) constituents. We stand to get an error, obviously, if we average over less than a complete cycle for any one constituent. As it turns out, there is no length of time that covers exactly a whole number of cycles for all the constituents present but we come pretty close for the major ones if we average over *25 hours*: this is roughly twice the M_2 tidal period (24.84 hours) and falls very near the average period for the main diurnal tidal constituents, K_1

and O_1 (23.93 and 25.82 hours, respectively). Simple averaging is the quick way to extract non-tidal currents but if we choose to, we can also extract them from a series of observed hourly currents using harmonic model predictions of the type described in Chapter 4.

Non-tidal currents, then, can include both density currents and wind-induced currents. Over longer periods of time (many tidal cycles) we can expect the wind currents in an estuary to 'average-out', leaving density currents as a variable but non-zero component of the flow regime. The organization of the density currents normally reveals an icon of the river estuary: its classic *two-layered circulation system*. In its ideal form, the two-layered system involves a net seaward flow at the surface and a net landward flow near the bottom with a layer having no net motion sandwiched in between. There can be many variations on the basic form, including cases where more than two layers are involved.

6.7 TIDAL AND NON-TIDAL FLUSHING IN ESTUARIES

One of the reasons for making a clear distinction between tidal and non-tidal currents is the recognition that the two can have very different effects on the transport of water-borne materials, including pollutants, in an estuary. This may be especially true in small sub-estuaries such as the Elizabeth River in Hampton Roads. This river unfortunately has a legacy of industrial contaminants present in its bottom sediments as well as a continuing influx through storm water runoff into its numerous small branches (Fig. 6.6). With high contaminant loadings always in the background, the effect of shoreline extensions and channel deepening on circulation and flushing ability become particularly important. Fig. 6.13 below shows where the currents come in.

As Fig. 6.13 illustrates, the ability of the sub-estuary to flush itself comes about through two modes. The first mode, **tidal flushing**, depends on the volume of water entering or leaving the system during each half of the tidal cycle. The average volume, or *tidal prism*, can be calculated as the product of the tidal range times the waterway surface area. In addition to these variables, tidal flushing also depends on the degree of mixing between resident waters and the water entering or leaving through the river mouth. In this mode, a net amount of contaminated material can be transported out of the estuary even if the net volume of water entering and leaving is zero.

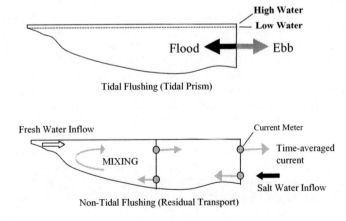

Fig. 6.13. Two types of flushing in an estuary.

The second mode, **non-tidal flushing**, depends not only on the degree of mixing but also on two other variables: the volume of fresh water entering at the head and the volume of salt water entering at the mouth of the system. Instead of the tidal prism, the key feature here is the **residual** or *net transport* at various depths, a feature that can be likened to a conveyor belt slowly but steadily moving material in at the bottom and out at the surface.

Contemplating the changes that might occur in their variables, it is easy to see that the two flushing modes just described can change in quite different ways. Reducing the waterway surface area by extending the shoreline or infilling, for example, can obviously reduce the tidal prism and tidal flushing. A more subtle change - deepening the entrance channel of the river - may enhance salt water inflow at the bottom and thus increase the residual transport and non-tidal flushing. Modeling studies are required to determine what kind of change – and how much – is likely to result from a development project involving harbor modification. In this way, the effects of the port development work can be evaluated well *before* the project is built.

Residual current example – At the conclusion of Chapter 4, a general formula was presented for predicting both tides and tidal currents. In this chapter the concept of a non-tidal or 'tidally-averaged' current was introduced. Averaging over multiple tidal periods or 'low-pass' filtering of a time series of observed current is one way of getting a look at the non-tidal current at longer intervals (intervals much longer than the tidal period). Another way to view the non-tidal current over short intervals of time makes use of the following mathematical model that retains the predictive element:

$$U'_t = U_t - U(t) \tag{6.2}$$

where

U'_t = zero-average non-tidal current at times $t_1, t_2, \ldots, t_n$
U_t = time series of current measurements at times $t_1, t_2, \ldots, t_n$
$U(t)$ = tidal current predicted at times $t_1, t_2, \ldots, t_n$

The success of the above model depends on a successful prediction of the *astronomical tidal current* using the right mix of the tidal harmonic constituents as described in Chapter 4. If an important constituent is left out of the formula, or is not represented by the right combination of amplitude and phase in calculating $U(t)$, it then winds up as part of U'_t. The current given by eq. (6.2) is sometimes called the *residual current* or the current 'left-over' after the predicted tidal current, including the mean, is removed. The separation of tidal and non-tidal currents is particularly useful in analyzing the output of three-dimensional hydrodynamic models: models that can be used to simulate a series of current measurements in navigation channels of heavily industrialized estuaries where measured currents are difficult to obtain. An example is available from model results, included wind effects, for the surface waters of the James River near Newport News Point in Hampton Roads, Virginia (Fig. 6.15).

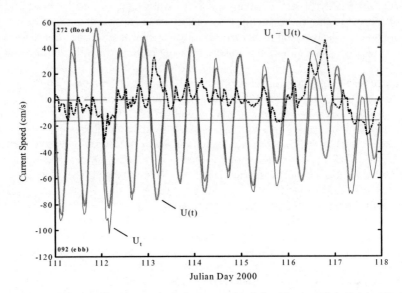

Fig. 6.15. Simulated surface current in the James River below Newport News Point. Flood and ebb current directions parallel the direction of the Newport News shipping channel in Hampton Roads. (Julian Day 2000 is the time in days starting 1 January 2000).

As illustrated in Fig. 6.15, the hydrodynamic model produced values of U_t every half hour for a seven-day period[3] (black curve). These were resolved into principal axis components with flood heading of 272° and ebb heading of 092°. The average current speed over 14 flood and ebb cycles during this period was thus –16.2 cm/s; i.e., 16.2 cm/s in the ebb direction (solid horizontal line). The current prediction formula from Chapter 4 was used to generate corresponding values of $U(t)$ representing the astronomical current with the average speed included (gray curve).

By including the mean value of the simulated current in $U(t)$, a clearer picture of the overall fit between $U(t)$ and U_t can be seen in Fig. 6.15. The residual current series (eq. 6.2) results from their difference (dash-dot curve). Of interest to the modelers was the fact that the simulated residual current briefly reached a maximum strength of 48 cm/s in the flood direction – a result consistent with strong easterly winds that occurred in Hampton Roads on April 25 (Julian Day 116), 2000. Otherwise, the most remarkable aspect of the residual curve is its *randomness*; its average is zero and, other than the inferred wind-induced, transient currents simulated by the hydrodynamic model, there are no obvious cycles with tidal or other periods hinting at further predictability on a long-term basis.

---

[3] Julian Day 111-117 or April 20-26, 2000.

7

Analyzing tides and currents

In Chapter 4, harmonic tidal constituents were introduced as the building blocks of the tide. Tidal types were identified based on the mix of harmonic amplitudes for four major tidal constituents represented by the symbols M_2, S_2, K_1 and O_1 and a model of the *astronomical tide* was presented using the amplitude and phase of these four along with other important constituents (constituents whose amplitudes in a given region are important contributors to the total variation in water level). But to take advantage of these concepts, we have to know how to extract a set of values for the harmonic tidal constituents from a record of water level or water current measurements. How is this done? This chapter will answer that question. And in the process, we will discover why tides and currents behave as they do in different water bodies around the world.

7.1 WHAT IS HARMONIC ANALYSIS?
Harmonic analysis is a mathematical technique applied to measured water levels and water currents that estimates amplitude and phase for a set of tidal constituents representing the astronomical tide. But there is an important 'given' in tidal harmonic analysis: *The constituents are uniquely identified by their frequencies.* Thanks to our knowledge of celestial mechanics, we know these in advance.

Unfortunately, when analyzing a finite set of measurements, we have no way of knowing the exact values of the amplitude and phase for the active constituents, or the *tidal harmonic constants* as they are called. We need to be especially careful here because the best we can come up with is an estimate of the tidal constants. Part of the reason is that the word 'constant' may be somewhat of an overstatement. Small changes in amplitude and phase occur with time, changing substantially in some cases because of man-made modifications in tidal waterways. In addition, water level measurements often contain non-tidal variations lumped together as 'weather tides' that must be carefully separated and distinguished from the astronomical tide. Another reason is statistical in nature and has to do with the fact that every record has a beginning and an end and contains measurement error; we are sampling a complex process that, for all practical purposes, has no beginning and no end. Unless our records are sufficiently long, interactions appear between constituents with similar frequencies that alter both their amplitude and phase (See Sec. 7.6).

What kind of record? - Statistical considerations suggest that we want a long record rather than a short one and that we should collect measurements at regular intervals spaced not too far apart. How long a record? Intuition tells us that it would be prudent to avoid sampling only a part of any of the *longer* cycles we can expect – the spring-neap cycle for instance. That suggests a sample at least 14 days in length and preferably a lunar month containing two spring-neap cycles (29 days rounded down to the nearest whole day). How short a sampling interval? In order not to 'under sample' an important *short* cycle, it is necessary to collect a measurement at least twice during a cycle of the tidal constituent with the shortest period[1]. At locations with enhanced shallow-water tides, that's likely to be the overtide M_8 with a period of 3.1 hours. Taking half that number and rounding downward, we'd want to sample not less than once every hour. But let's not forget measurement error and the fact that tide gauges, and current meters, are likely to experience quasi-periodic or random noise due to the residual effects of wind waves and ship wakes that manage to escape filtering devices like the stilling well on a tide gauge. Sampling every six minutes, or ten times an hour, makes better sense for tidal analysis. We can then reduce the noise by integrating (smoothing) the recorded values using a low-pass digital filter with high frequency cutoff (Appendix 1).

The least squares fit – Suppose we have a water level record consisting of smoothed measurements taken on the hour, 24 times a day, for 29 days – 696 measurements in all. We can get an estimate of the tidal constants for the astronomical tide in that record, or any record with *N* measurements, using a procedure called **Harmonic Analysis Method of Least Squares (HAMELS)**. Here's a brief explanation of HAMELS and how it works (a complete explanation is given in Chapter 10):

(1) A formula like that of Eq. (4.2) in Chapter 4 is set up with *m* harmonic constituents to predict the astronomical tide height, *h*, at any time *t*, relative to h_0, the mean water level in the series:

$$h(t) = h_0 + \sum_{j=1}^{m} H_j \cos(\omega_j t - \phi_j)$$

(2) Differences are formulated between the observed water level, h_t, and the predicted tide, *h(t)*, at each measurement time *t*. This gives *N* = 696 hourly differences; each one is then squared so that the resulting numbers are all positive before they are added together in a single sum. The *sum of squared differences* is expressed by the formula

$$SSQ = \sum^{N} [h_t - h(t)]^2 \qquad (7.1)$$

(3) The harmonic constants, H_j and ϕ_j, for each constituent with frequency ω_j are determined in such a way that *SSQ* becomes as small as possible for the record in hand and the set of frequencies (*j* = *1,m*) used to analyze it. This is the 'best fit' of the model to the data in the least squares sense. To judge how good the fit actually is, we will turn to some important statistical parameters, the *root-mean-square* (*RMS*) *difference* and the *variance* or *mean square value* (the average of the squared deviations about the mean of the *N* observations).

---

[1] If the sampling interval is greater than half the period of a sinusoidal wave, that wave will appear in your sample with a longer period than it actually has – a process known as *aliasing* - see Chapter 10, Sec. 10.2.

Sec. 7.1]   **What is harmonic analysis?**   95

The method of least squares is something to marvel at. For *SSQ* in eq. (7.1) to be zero, every one of the observed and predicted value pairs in the series would have to be identical. In a real world with measurement error, that never happens. But the answers that come out of step (3) above are unequivocal: For each water level series and set of harmonic constituents chosen, there is a minimum *SSQ* number and a unique set of harmonic constants for that series. No other combination of numbers producing a lower *SSQ* is possible. Still, this does not mean that the analysis has necessarily produced the best set of tidal constants for a given location. Another sample will no doubt produced slightly different results and sampling theory suggests that either a longer record or sample averages (*vector averages* of the amplitude and phase) may be required to arrive at the best solution. Certain constituents have amplitudes and phases that, if sampled monthly, show systematic variation over a cycle of half a year (see Sec. 7.?).

Taking the last point a bit further, the ideal record length for a complete tidal analysis is 18.6 years (the lunar nodal cycle) but a record of one or more years allows a considerable number of tidal constituents to be determined accurately. The name and origin of the additional constituents is not important here but many have frequencies close to those of other constituents and a longer record is therefore required to resolve them[2]. However, a complete one-year water level record - to say nothing of a one-year water current record – is not easy to obtain without considerable effort and expense, keeping quality control in mind. Useful information can be gained from records one to six months in length and certain statistical measures help us judge their quality.

Statistics you can use – To judge how well any single tide prediction is likely to agree with the observed water level at a place, the *Root-Mean-Square difference* or **RMS error** is a useful index. We get this number by taking the square root of *SSQ* divided by the number of measurements ($N = 696$ in the above example). If a certain water level analysis should yield an *RMS* error of, say, 10 cm, it would infer an agreement within ± 10 cm roughly two-thirds of the time, assuming a normal distribution of differences. Another useful statistic is the **Reduction in Variance (RV)** achieved by the tide model used in the analysis. It is calculated as the ratio of the model variance to the total variance of the series, or

$$RV = \frac{\sum [h(t) - h_0]^2}{\sum [h_t - h_0]^2} \qquad (7.2)$$

where both summations are made for the *N* sample times. For example, $RV = 0.90$ would indicate that 90 percent of the observed variance in water level about the series mean, h_0, is accounted for by the tide model. This would leave only 10 percent as variance unaccounted for after applying the 'best fit' prediction model. In addition to statistical parameters, we should also look at the **residual curve** - a plot of the difference between the observed and predicted tide. It takes a careful eye to spot them but both tidal and non-tidal variations that the model fails to capture in the record being analyzed will show up in the residual curve. Special techniques such as a **Fourier analysis** of the residual curve can help to identify them.

---

[2] For example, the lunar-solar semidiurnal constituent K_2 (30.0821deg/hr) requires at least 6 months of record to resolve from S_2 (30.0000 deg/hr). See Sec. 7.6.

7.2 HARMONIC ANALYSIS OF WATER LEVELS: *SIMPLY TIDES*

Least squares harmonic analysis of tides can perhaps best be demonstrated through example. In preparing my examples, I developed a program for the personal computer called *SIMPLY TIDES* that employs a simple Graphical User Interface (GUI) and the MATLAB® programming language to perform both analysis and prediction of tides. Because this really is a 'hands-on' book, the next two paragraphs will include a web site introduction and the specific information you need to do both. However, if you do not have access to MATLAB, just skip to the examples in Sec. 7.3 below.

Copies of the program *SIMPLY TIDES* and a tutorial describing its use are available for downloading from the Physical Sciences research web site at the Virginia Institute of Marine Science (http://www.vims.edu/physical/research/) or from the MATLAB Central File Exchange (http://www.mathworks.com/matlabcentral/). To ensure a plentiful supply of interesting water level data from online sources around the world, *SIMPLY TIDES* is designed to use data direct from government web sites such as the Center for Operational Oceanographic Products and Services (CO-OPS), a unit of the US National Ocean Service (http://co-ops.nos.noaa.gov), and the National Tidal & Sea Level Facility of the British Oceanographic Data Centre (http://www.pol.ac.uk/ntslf/). To download data from CO-OPS, for example, just follow these two steps:

Access data - On reaching the *CO-OPS* homepage, select *Water Level Observations* to access NOAA's extensive collection of verified historical water level data at US and Global Coastal Stations. After choosing a station, data can be requested in several formats but the standard choice is hourly heights referred to the *MLLW* datum using *local standard time (LST)* and the default date and time format (*yyyy/mm/dd hh:mm*). After pressing the *View Data* button, the requested data then appear onscreen where you can verify that at least 29 days of recorded hourly heights[3] are available.

Transfer data - Transfer the data using the WINDOWS 'copy' and 'paste' commands directly into a Microsoft *EXCEL* workbook. After a 'paste' into column A of the first worksheet, use the '*Data / Text-to-Columns*' command to *parse* the data from text into numeric data in separate columns. Text files of water level data obtained from the British Oceanographic Data Centre, U.K. National Tidal & Sea Level Facility, can be transferred and parsed using the same format. It's useful to copy station header information into the second page (second worksheet) for reference purposes. Water level data from other sources can also be entered into the Excel workbook using a simple three-column format: Station ID (or sample number), Excel date & time (date format type *3/14/98 13:30*), and water level (meters or feet).

Why Excel? Using the *xlsread* function, MATLAB reads input data directly from the first page of an Excel workbook with filename extension *.xls*. Following the execution of this function, all of the numeric data in the file are placed in an n x m matrix ready for use by *SIMPLY TIDES*. The key reason for choosing Excel is the handling of serial date and time. Although you see a date and time in a date-formatted cell (e.g., *3/14/98 13:30*), MATLAB sees a single number (35868.56250) representing the fractional number of days since midnight beginning December 31, 1899. For convenience, days of the year can be indexed as **Julian days (1-366)** in graphic data plots. While the

---

[3] NOS water levels recorded at 6-minute intervals are integrated to yield hourly heights.

time origin used by Excel is arbitrary, it is nevertheless extremely convenient for time series calculations representing different years and even different centuries (remember the 'Y2K' problem?). And Excel makes it easy to form a time column in your time series array. For example, given a starting date and time, you can enter a formula in the cell below that adds a fixed time increment to this date (e.g., 1/24 for hourly incrementing), copying the same formula into succeeding rows further down. Building a serial date & time column next to a column of dates and times provided by your data source is an excellent way to check the latter for 'drop-outs' and other timing errors.

7.3 SOME EXAMPLES OF WATER LEVEL ANALYSIS

In the examples that follow, we will do more than simply derive the set of harmonic constants that best fits a given water level history. We will be interested in knowing if the fit, in terms of the variance accounted for, was less than perfect (it will be) and what is the reason for it? Aside from measurement error, there are two possibilities: 1) One or more components of the 'true' astronomical tide are missing from the model, 2) variations in water level are present that have a non-tidal, weather related origin In order to make better predictions, we're interested in learning as much as we can about both possibilities and will look for the answers in the *residual curve*.

Chesapeake Bay Entrance, US East Coast

NOS tide station no. 8638863, CHESAPEAKE BAY BRIDGE TUNNEL, VA, lies at the Chesapeake Bay entrance to the North Atlantic Ocean. It's a good example of a station with an abundance of data for all seasons of the year that can be easily downloaded from the net. To demonstrate, we will use a 29-day record starting September 1, 2002 and analyze it using *SIMPLY TIDES* to fit nine tidal constituents (M_2, S_2, N_2, K_1, O_1, M_4, M_6, S_4, MS_4). The results appear in Fig. 7.1.

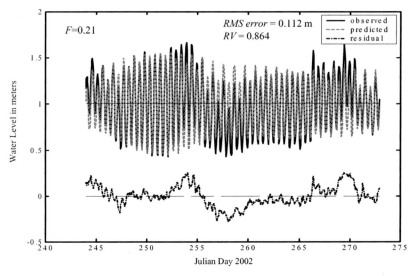

Fig. 7.1. Harmonic analysis of 696 hourly heights at the Chesapeake Bay Bridge Tunnel (CBBT) Virginia starting September 1, 2002 (JD244).

Selection of the nine major tidal constituents for *SIMPLY TIDES*, the tidal model applied in Fig. 7.1, was guided in part by prior knowledge of constituent amplitudes in the Chesapeake Bay region but also by the relatively large separation in their frequencies within the main semidiurnal, diurnal, and quarter-diurnal tidal types (see Table 4.2). This combination is a general one that can be used in many regions of the world to reveal the basic composition of the tide. Obviously a more complete representation with many additional constituents (together with the longer records needed to obtain them) is required for optimum tide prediction capability. Meanwhile, let's see what we can learn from these nine.

After an initial glance at Fig. 7.1, it's clear that the fitted or model-predicted heights (dashed curve) do not match the observed heights (solid curve) shown in the figure. Although we have a relatively small error (*RMS error* = 0.11 m) and fairly large reduction in variance (*RV* = 0.864), a glance at the *residual* (dash-dot curve) reveals some irregular (quasi-periodic) sub-tidal variations in water level taking place during this 29-day interval. The term sub-tidal refers to water level oscillations at frequencies less than about 1 cycle per day. A history of the sub-tidal variations in the signal could also be obtained by applying a low-pass digital filter with low-frequency cutoff to strip away oscillations at tidal and higher frequencies. In practice this is not always possible because such a filter must have a very steep response curve near the minimum tidal frequency of 1 cycle per day; otherwise the 'passed' signal immediately below that frequency will contain only a partial response (think chopped-off peaks). It would also defeat the purpose of the present analysis: to judge the ability of the astronomical tide model as configured to reproduce observed variations in water level.

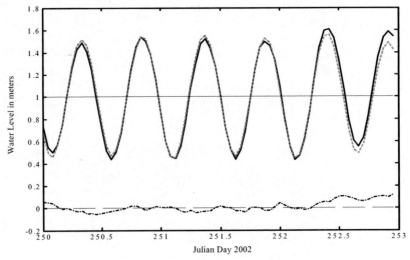

Fig. 7.2. Series of 72 hourly heights at Chesapeake Bay Bridge Tunnel (CBBT) starting September 7, 2002 (JD250). See legend in Fig. 7.1.

To confirm the adequacy of the tide model we should carefully inspect the residual curve at a reduced time scale for any remaining oscillations at tidal frequencies. A 3-

day plot feature of *SIMPLY TIDES* assists in this effort. The 3-day plot in Fig. 7.2 offers visual assurance that no significant oscillations at known diurnal or semidiurnal frequencies are unaccounted for; otherwise they would appear in the residual. It's also reassuring to note the close agreement in both amplitude and phase of the observed and predicted curves in this time segment. Finally, the twice-daily oscillations appearing in the figure and the *tidal form number* (F=0.21), a parameter introduced in Chapter 4 (Sec. 4.3), confirm that the type of tide is fully semidiurnal.

More about sub-tidal variations in water level - Research by oceanographers in the mid-to-late seventies revealed that sub-tidal sea level variations at periods between 3 and 5 days are common events along the US East Coast, being particularly evident in winter. Analyzing a second record from the Chesapeake Bay Bridge Tunnel for November 2002, we see a very strong non-tidal oscillation with a period of 4.8 days (Fig. 7.3) and we note a further drop in the RV value to 0.723. Although still a transient oscillation, this feature persisted throughout November and most of December as well, reaching amplitudes of almost 0.5 m at times. Clearly these are fluctuations on a par with the tidal signal itself, greatly impacting the predictability of the total tide (astronomical plus meteorological) during sub-tidal events. Virtually the same residual as shown in Fig. 7.3 appeared simultaneously at stations on the outer coast (e.g., Duck, North Carolina) as well as inside Chesapeake Bay and its lower tributaries including the York River (e.g., Gloucester Point, Virginia). They indicate a coupled coastal ocean - estuarine response to cyclical wind systems associated with winter storms (extratropical cyclones or 'northeasters'); storms so broad they usually cover several states at once.

For sub-tidal oscillations with periods longer than three days, low-pass digital filters are effective if a smooth curve is wanted: compare the residual curve in Fig. 7.3 with the solid line derived by applying a 36-hour LP filter (see Appendix 1).

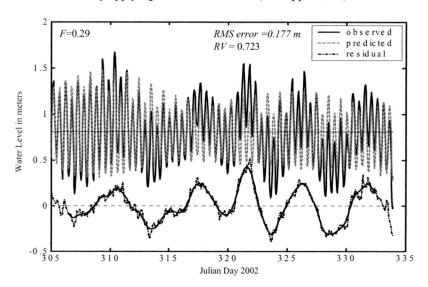

Fig. 7.3. Harmonic analysis of 696 hourly heights at the Chesapeake Bay Bridge Tunnel (CBBT) Virginia starting November 1, 2002 (JD305). Solid line through the residual curve is the low-pass filtered residual water level.

The period of 4.8 days for the residual curve in Fig. 7.3 was obtained using another tool included in *SIMPLY TIDES*: the *residual periodogram* or *line spectrum*. The periodogram displays the raw output of a Fourier analysis (using MATLAB's Fast Fourier Transform or FFT) as shown in Fig. 7.4. Although the resulting energy peaks are not without error due to a lack of smoothing of the raw spectral estimates, the associated frequencies in cycles per day are accurately represented, which is the information most needed to characterize the residual in this example. A description of Fourier analysis, its similarities and differences from harmonic analysis is given in Chapter 10, Sec. 10.2.

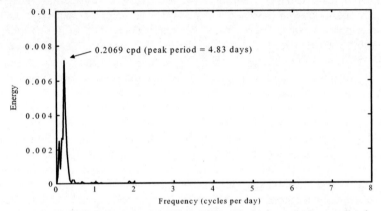

Fig. 7.4. Fourier periodogram of the model residual (observed minus predicted water level), Chesapeake Bay Bridge Tunnel (CBBT) Virginia, November, 2002.

River Mersey, West Coast of England

The River Mersey is a shallow estuary opening into Liverpool Bay and the Irish Sea. We will now examine a 'sea level' record, as they are called in the United Kingdom, from the Liverpool sea level station on the Mersey, one of 44 stations in the UK National Network (Fig. 7.5) operated by the Proudman Oceanographic Laboratory for the UK National Tidal and Sea Level Facility (NTSLF). As noted previously in Chapter 4 (Sec.4.3), the US NOS have chosen to use the term 'water level' when referring to such measurements while the UK NTSLF prefer the term 'sea level'. NTSLF makes their data freely available for downloading but asks that you register first so that they have a record of data usefulness. Similar to the US NOS water level data, UK NTSLF sea level data are easy to access and transfer from their web site and are fully compatible with the Excel time and date format used by *SIMPLY TIDES*.

Fig. 7.5. National Network of UK tide stations.

Sea levels for Liverpool provide us with a classic example of fully semidiurnal tide conditions - as underscored by a tidal form number of only 0.07 for the record shown in Fig. 7.6. The range of tide at Liverpool is obviously much larger than in Chesapeake Bay, reaching more than 8 meters during the perigean-spring tides displayed here. That is why we should not be surprised to find a much larger reduction in variance (RV = 0.995) using the same 9-constituent tide model during the fair weather conditions that seem to have prevailed at Liverpool in June 2002. We note, however, that the RMS error (Fig. 7.6, 0.179 m) is about the same as the one representing Chesapeake Bay fall/winter conditions (Fig. 7.3, 0.177 m); this suggests that the tide model can still be improved, even though it already accounts for more than 99 percent of the total variance in sea level during this particular month at Liverpool.

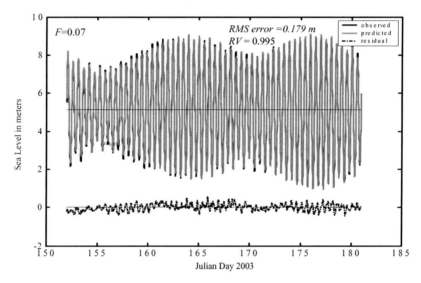

Fig. 7.6. Harmonic analysis of 2784 15-minute heights from Gladstone Dock, Liverpool (GDL) England, starting June 1, 2002 (JD152). Heights refer to Admiralty Chard Datum (ACD) of Lowest Astronomical Tide (LAT).

Looking closely at the residual in Fig. 7.6 we notice that while it is relatively flat, suggesting the absence of any significant sub-tidal oscillations, it has persistent low-amplitude fluctuations at what appear to be tidal frequencies. Here again the Fourier periodogram can be of use in confirming this interpretation. A word of caution first: There is an inherent limitation in the periodogram imposed by the Fourier frequencies, the set of frequencies defining a Fourier line spectrum, each of which is a multiple of the **fundamental frequency,** the inverse of the sampling duration (29 days in this case). With the exception of the overtides, the astronomical tide constituents do not have frequencies that are exact multiples of one another and thus will not coincide, as a group, with any set of Fourier frequencies. This is the reason why we're using HAMELS, and not Fourier analysis, to analyze tides in the first place. But we can still obtain clues of a missing constituent whenever significant variance (energy) appears in the neighborhood of a known tidal frequency. This is exactly what we see in the periodogram for the Liverpool example as shown in Fig. 7.7.

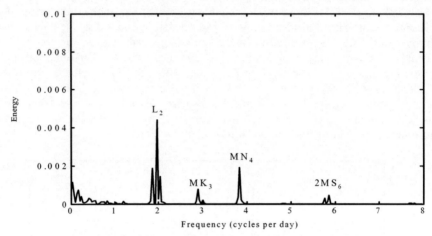

Fig. 7.7. Fourier periodogram of the model residual (observed minus predicted water level), Gladstone Dock, Liverpool (GDL) England, June, 2002.

The largest peak in Fig. 7.7 belongs to a lesser cousin of the constituent N_2 known as the smaller lunar elliptic semidiurnal constituent: L_2 (speed: 29.528 deg/hr). A few of the additional semidiurnal constituents that our basic tide model does not include cluster around it. Since the tidal type at Liverpool is so strongly semidiurnal, it would be useful to include L_2, which, though normally a minor constituent, might add more than a centimeter to the predicted tide at this location. Appearing at higher frequencies are some examples of shallow-water tides (MK_3, MN_4, $2MS_6$) that arise here due to the shallowness of the depth relative to the tide range in the approaches to the River Mersey.

Table 7.1 contains a listing of the nine constituent amplitudes for the September analysis at the Chesapeake Bay Bridge Tunnel (CBBT, Fig. 7.1) and the June analysis at Gladstone Dock, Liverpool (GDL, Fig. 7.6) in 2002 along with their ratios to the dominant M_2 amplitude at each location. We see that while the diurnal amplitudes (K_1, O_1) are about three times larger at GDL than at CBBT, their ratios with M_2 are almost three times smaller in these samples. This is due to the much higher amplification of M_2 at GDL: by a factor of more than eight compared to CBBT. Notice that the amplitude of M_4, the first overtide of M_2, is about thirty-seven times greater at GDL than at CBBT. This evidence points to the considerable importance of shallow-water effects on the M_2 tide, its overtides (M_4, M_6) and compound tides (MS_4, MN_4) at Liverpool.

Table 7.1. Tidal amplitudes (m) at CBBT: Sep2002 (row 1); GDL: Jun2002 (row 3).
Shaded cells contain ratios of constituent amplitudes to M2 amplitude.

M_2	S_2	N_2	K_1	O_1	M_4	M_6	S_4	MS_4
0.373	0.084	0.106	0.052	0.039	0.006	0.006	0.003	0.005
1.00	0.22	0.29	0.14	0.10	0.02	0.02	0.01	0.02
3.073	0.689	0.487	0.150	0.118	0.226	0.055	0.011	0.094
1.00	0.22	0.16	0.05	0.04	0.07	0.02	0.00	0.03

Northwest Persian Gulf

The Persian Gulf (Arabian Gulf) is a marginal sea with an average depth of 35 m, an average width of 200 km, and an axial length of 800 km, or roughly three and a half times the dimensions of Chesapeake Bay. As in Chesapeake Bay, the tides of the Gulf derive from an external source (Gulf of Oman and the Arabian Sea) and propagate into it through the narrow Straits of Hormuz.

The dimensions of the Persian Gulf are large enough to ensure that the *Coriolis effect* has a major influence on water movements - thus Kelvin waves can be expected to enter the picture. In fact, the Gulf allows multiple amphidromic systems to operate within its bounds, systems that may be grouped according to tidal type: diurnal and semidiurnal. These two groups are different yet equally important, making a single set of co-tidal and co-range charts (as presented in Chapter 5 for Chesapeake Bay) impractical to use in the Gulf. To better understand how tides function in the Persian Gulf, we can make good use of constituent amplitude and phase diagrams as presented in a Co-tidal Atlas developed by the U.K. Hydrographic Office. After separating the major tidal constituents through harmonic analysis, each constituent is treated as a separate waveform that behaves as depicted in the diagram. Two of these diagrams are shown on the next two pages to illustrate the more complex behavior of tidal systems that have reached the amphidromic stage.

Fig. 7.8 represents the major semidiurnal tidal constituent M_2, or the M_2 tide, as we will call it. It rotates about a pair of amphidromic points, one visible in the northern Gulf, the other appearing in the southern Gulf just outside the lower right corner of Fig. 7.8. Fig. 7.9 shows one of the main diurnal tidal constituents, the K_1 tide, and we note that it has only one amphidromic point located near the center of the Gulf between the M_2 amphidromic points. Diagrams for the S_2 and O_1 tides are similar in appearance to M_2 and K_1, respectively, and are not shown here.

A zone of minimum amplitude for the M_2 tide is indicated in Fig. 7.8 by an amphidromic point lying a short distance offshore from the port of Safaniya in the Eastern Province of Saudi Arabia, just below the Kuwaiti border. Safaniya lies well to the northwest of a similar amphidromic point for the K_1 tide near the port of Ras Tanura as shown in Fig. 7.9. This is strong evidence that the tidal type will progressively shift from semidiurnal to nearly diurnal when moving in a northwesterly direction between Ras Tanura and Safaniya. We will check this assumption by analyzing a water level record from each location (data provided by Coastal Environmental Associates, Inc., consultants to the Arabian American Oil company).

Ras Tanura tides - Analysis of a 29-day record of hourly heights at Ras Tanura fully confirms its semidiurnal tidal type. One sign of this is the pronounced *spring-neap cycle* shown in Fig. 7.10. The tidal form number for this station is $F=0.21$, significantly higher than the value for Liverpool, England, but approximately the same as the value calculated for the Chesapeake Bay Bridge Tunnel. Over 90 percent of the variance in this record from Ras Tanura is accounted for by the tide model ($RV = 0.91$) and the *RMS* error is 0.16 m. The close match between the 3-day plot curves shown in Fig. 7.11 completes the picture of a good fit. No surprises here, with the possible exception of the residual in Fig. 7.10. It shows a sub-tidal oscillation with amplitude of nearly half a meter and a period of roughly four days. We will compare this oscillation with a corresponding one at Safaniya farther up the coast.

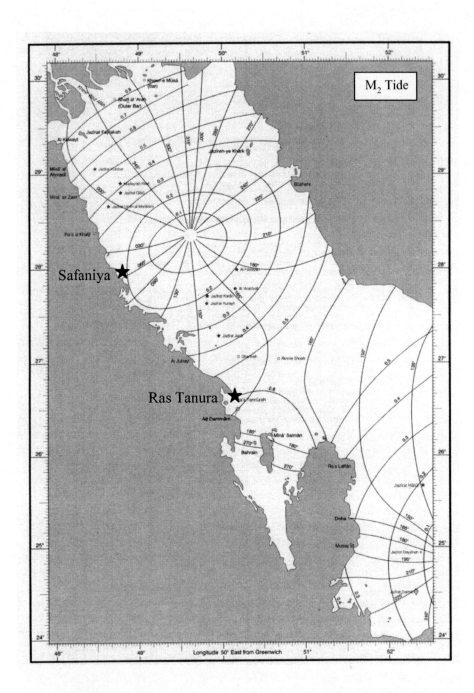

Fig. 7.8. Amplitude and phase for the M_2 tide in the Persian Gulf (reproduced from Admiralty Co-Tidal Atlas of the Persian Gulf, Pub. NP214, Edition 2, 1999 with permission; United Kingdom Hydrographic Office, Taunton, England).

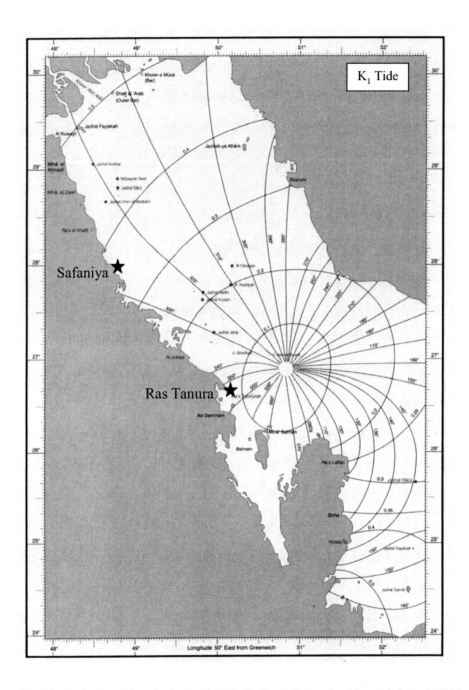

Fig. 7.9. Amplitude and phase for the K_1 tide in the Persian Gulf (reproduced from Admiralty Co-Tidal Atlas of the Persian Gulf, Pub. NP214, Edition 2, 1999 with permission; United Kingdom Hydrographic Office, Taunton, England).

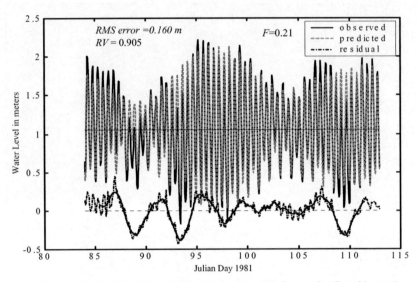

Fig. 7.10. Harmonic analysis of 696 hourly heights at Ras Tanura, Saudi Arabia, starting March 25, 1981 (JD84). Solid line through the residual curve is the low-pass filtered residual water level.

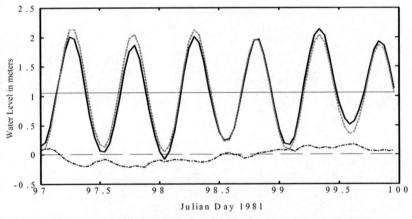

Fig. 7.11. Plot of 72 hourly heights at Ras Tanura starting April 7, 1981 (JD97).

Safaniya tides – The port of Safaniya lies about 200 km northwest of Ras Tanura on the east coast of Saudi Arabia. In the space of this relatively short distance the tidal type changes dramatically from semidiurnal to mostly diurnal, as underscored by a form number of $F=1.6$ ($F=1.5-3.0$ is mixed, mainly diurnal). Fig. 7.12 shows the tidal variation at Safaniya over the same 29-day time period used in the Ras Tanura analysis. In place of the spring-neap cycle observed at Ras Tanura (Fig. 7.5), the range variation at Safaniya is dominated by the *tropic-equatorial cycle*, the foremost signature of the diurnal tide. This fact leads to other differences between these two stations.

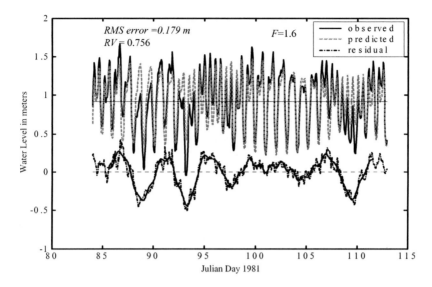

Fig. 7.12. Harmonic analysis of 696 hourly heights at Ras Tanura, Saudi Arabia, starting March 25, 1981 (JD84). Solid line through the residual curve is the low-pass filtered residual water level.

Because the tropic month (27.3 days) is shorter than the lunar month (29.5 days), tropic tides will progressively lag behind spring tides by a little more than two days every month. For that reason, it's no surprise to find tides of maximum range on entirely different days at Ras Tanura and Safaniya; while tides of maximum range occur on Julian days 97 and 111 at Ras Tanura, they occur on Julian days 89 and 102 at Safaniya. But the Safaniya tides are not entirely diurnal. A series of three-day plots at Safaniya (Figures 7.8 - 7.10) depicts a typical monthly progression as follows:

1. semidiurnal tides of minimum range (days 94 to 97)
2. mixed tides of intermediate range (days 97 to 100)
3. diurnal tides of maximum range (days 100 to 103).

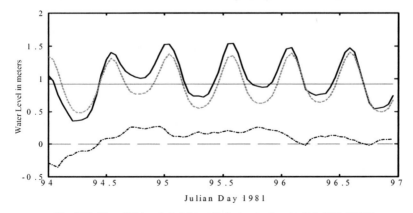

Fig. 7.13. Plot of 72 hourly heights at Safaniya starting April 4, 1981 (JD94).

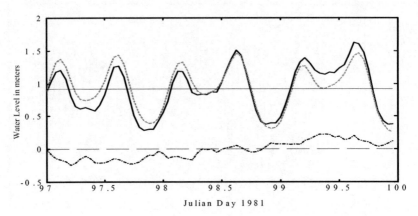

Fig. 7.14. Plot of 72 hourly heights at Safaniya starting April 7, 1981 (JD97).

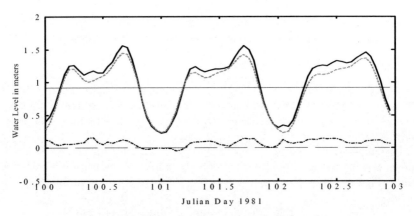

Fig. 7.15. Plot of 72 hourly heights at Safaniya starting April 10, 1981 (JD100).

Comparing the residual signals for Ras Tanura and Safaniya (Figs. 7.10 and 7.12), you'll note that they are virtually identical. Because the observed tidal range at Safaniya is clearly less than at Ras Tanura, the reduction in variance achieved by the tide model is smaller there ($RV = 0.76$). This is not an indication that the model is less successful in predicting the *astronomical tide* at Safaniya, rather it underscores the fact that the relative contribution of the *weather tide* is greater at this station. The residual curves in Figs. 7.10 and 7.12 are another example of a *sub-tidal oscillation* arising from the direct action of wind stress and pressure changes associated with large-scale weather systems. These appear in the northern Persian Gulf in connection with the *Shamal*, a seasonal meteorologic event in the region noted for very high winds lasting several days. During such events, forces of the atmosphere act over time and distance scales that are much larger than those governing the tides propagating within the Gulf. What is the take-home message? Two dissimilar processes are at work. Although the tidal characteristics were quite different for the two adjacent stations, the peak period of the residual oscillation during this time was 4.1 days at both Ras Tanura and Safaniya.

7.4 HARMONIC ANALYSIS OF WATER CURRENTS: *SIMPLY CURRENTS*

The good news here is that harmonic analysis of water currents is easy – the methodology is virtually the same as that used to analyze water levels. The bad news is that good quality current data are much harder to find. Unlike a tide station that can be conveniently installed on any shore-connected pier, a *current station* is normally installed away from shore at points where the currents are fully developed and free from the influence of structures. The instruments needed to make such measurements are both sophisticated and expensive. Oceanographers deploy strings of current meters beneath buoys, or use the latest device, a multi-beam **acoustic Doppler profiler (ADP)** or **acoustic Doppler current profiler (ADCP)** - to study estuarine circulation in three-dimensional space. This presents a choice between spatial and temporal sampling; ADP systems mounted on ships can cover a large region while sampling underway but with limited ability to repeat a specific sampling pattern at regular intervals of time.

From the US National Ocean Service's perspective, navigation channels are where the action is and their surveys emphasize measurements of long duration there (just as NOS nautical chart surveys always cover a shoal in considerable detail but don't waste many soundings on deep holes). Unfortunately current fields, like shoal depths, tend to change with time. For example, the *NOS Tidal Current Tables 2001 for the Atlantic Coast of North America*[4] contain broadly revised current predictions for lower Chesapeake Bay along with a disclaimer concerning earlier current data – data previously considered accurate but now held to be inaccurate due to dredging and bridge/tunnel construction. Daunting words but let's first introduce the harmonic method of current analysis and then we'll consider where to find the data we need.

Current dimensions - one, two, or three? – The most obvious difference between water level and water current measurements is that the latter involve at least two dimensions and sometimes three. We measure *horizontal currents* either in terms of speed and direction, or as orthogonal vector components designated U and V. An **electromagnetic current meter (ECM)**, for example, measures U and V directly and uses compass headings to computationally orient the axes *north-south* and *east-west*. A positive U reading then represents *eastward* flow and positive V represents *northward* flow with negative readings representing flow in the opposite direction (*west* or *south*). In some instances, for example near a *tidal front* in an estuary, a small vertical current may exist but tidal current surveys usually ignore this component.

The question is, can another dimension be dispensed with as well? In many cases the answer is yes. A numerical technique called **Principal Components Analysis (PCA)** tells when and how. The details of this technique are explained in Chapter 10 but its use is very straightforward and will be demonstrated here by example. We're looking for a directional reference for horizontal currents called the **principal axis** (principal component); we will project our U,V current readings onto this axis and simultaneously project them onto a **secondary axis** perpendicular to the principal axis. After we do this we will have a new set of readings: U_p, V_p. The **total variance** - the variance for U_p and V_p combined - will be the same as for U and V combined. However, thanks to PCA, U_p will have the greatest fraction possible of the total variance and V_p will have the least. If the U_p fraction is high enough, we may choose it and ignore V_p altogether, reducing the current's dimensionality from two to one. This often works well but not in every case.

---

[4] Presently published by International Marine, Camden, Maine.

7.5 SOME EXAMPLES OF WATER CURRENT ANALYSIS

San Francisco Bay -To demonstrate how the principal axis is used to obtain a single variable for a tidal current analysis, let's look at an NOS data set for water current available at Richmond, California, in San Francisco Bay (Fig. 7.16). This station, by the way, is one of the few in the United States currently offering continuous current records of adequate length for harmonic current analysis. Many field deployments obtain less than a full 29-day record; a choice can be made in that case to analyze a shorter record of 14-days, the minimum that will yield practical results.

From the NOS/PORTS[5] data archive http://co-ops.nos.noaa.gov, we will analyze a 14-day ADP record from the Richmond station starting November 2, 2002, with current speed in knots and current direction in degrees true. The currents were recorded near the surface at 6-minute intervals.

Fig. 7.16. San Francisco Bay PORTS stations (courtesy of the US National Ocean Service, NOAA).

Program *SIMPLY CURRENTS* first converts current speed and direction to U and V components with their means removed. The U,V current plot in Fig. 7.17 shows how the Richmond data are distributed (note the *bivariate mean* of the U,V data is near zero). Each gray dot in the diagram represents the tip of a current vector with speed and direction from the origin plotted on a grid with orthogonal U and V axes. Using PCA, we project the current vectors (all 3143 of them) onto orthogonal axes U_p and V_p. This action is the same as a *rotation* of U and V while keeping the dots fixed: the origin is first shifted to the bivariate mean position and both axes are then rotated 61° clockwise about this point. Thus a point lying on the U axis would have a current heading of 090° while a point on the U_p axis would have a heading of 151°. Examining the data and how they are arrayed from one end of the U_p axis to the other, it's apparent that the current readings projected onto this axis show considerable spread. In fact, the axis shown with reciprocal headings of 331° and 151° relative to true north delivers the greatest spread, the maximum variance possible for any axis orientation and U_p therefore becomes the principal axis in this example. Meanwhile V_p, the axis perpendicular to U_p, clearly has the least spread in projected values and it becomes the secondary or minor axis. Expressed as a fraction of the total variance, the *Principal Axis Variance (PAV)* in Fig. 7.17 is 0.992. The small remaining fraction of variance associated with the minor axis (0.008) suggests we wouldn't miss much if we simply ignored V_p and concentrated on U_p values instead. Note that the bivariate mean (the mean of U and V over the 14 days) can be added back to the data after axis rotation. This simple translation of the origin does not affect the distribution of U and V.

---

[5] Physical Oceanographic Real-Time System (PORTS)

Sec. 7.5] **Some examples of water current analysis** 111

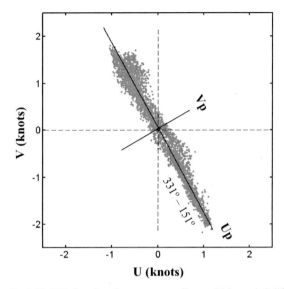

Fig. 7.17. U,V plot of surface current readings at Richmond, California, Nov.2002. Principal Axis Variance (PAV) is 0.992. Flood direction is $331°$; Ebb direction is $151°$.

Overlooking the mathematics involved in converting current readings measured along the U axis into current readings along U_p, we can see visually that they amount to nothing more than axis translation and rotation (with the dots 'frozen' where they are) until U captures the maximum variance possible in the U,V data set and thus becomes U_p. This is very much like the least squares procedure described earlier in the chapter with the 'squares' now directed to go the other way – toward the maximum value possible. Variance, also known as the *mean square value*, can be expressed here in the form

$$MSQ = \frac{1}{N}\sum_{}^{N} U_t^{2}$$

where U_t is the current speed at time t measured along the rotated U axis, N being the sample size (3143 in the above example); U_t has both positive and negative values and a zero mean.

Reciprocal headings like $331°$ and $151°$ for the principal axis leave us with one other problem that no algorithm can solve. The person doing current analysis must decide which of these headings corresponds to the *flood* direction and which corresponds to the *ebb* direction based on his or her knowledge of the location. Flood is generally taken to mean the landward direction with positive U_p representing flow in that direction and negative U_p representing flow in the seaward (ebb) direction.

The U_p current component is the one we want for tidal current analysis, not only because of its 'efficiency' in terms of explained variance but for another more practical reason: We can now analyze a **current curve** – current plotted as a continuous function of time – in exactly the same way that we analyzed water levels. Equally important, we will be able to examine a residual curve for the current.

Fig. 7.18 shows the results for an analysis of U_p using the harmonic model previously applied to water levels and the same set of tidal constituents: M_2, S_2, N_2, K_1, O_1, M_4, M_6, S_4, and MS_4.

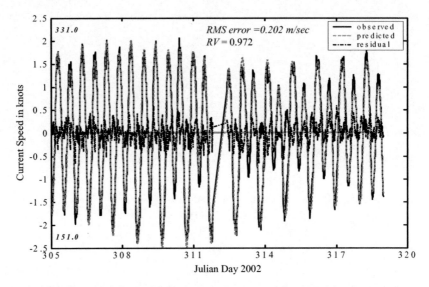

Fig. 7.18. 14-day harmonic analysis of principal axis surface currents at Richmond, California, starting November 1, 2002 (JD305).

Although there are a few individual 6-minute gaps spread across the record in Fig. 7.12, the pronounced data gap of about 15 hours on Julian days 311 and 312 is hard to miss. While we'd like to get 100 percent data recovery, it's especially unlikely when working with currents. But in spite of the missing hours, we're still able to extract a useful set of harmonic constants because the fit to the remaining data is unaffected by the gap. Statistical measures show that this 14-day analysis produces an RMS error of 0.20 knot and accounts for more than 97 percent of the variance in U_p.

Bi-directional plots tend to become a bit crowded. In the 29-day analyses presented below we will omit displaying the observed current to better distinguish the residual current – the main feature of interest in these plot diagrams.

Chesapeake Bay Entrance
While the goal of a fully operational NOS PORTS system in every major estuary across the US is an excellent one, recent federal funding limits have had their impact. All five pages of *Important Notices* published in the NOS *Tidal Current Tables 2003* underscore that fact, citing the "extremely high cost of new instrumentation" as well as the "cost of verifying the accuracy of presently published data". As a result, most US estuaries now lack an NOS PORTS system offering either real time or historical current data, including the PORTS system in Chesapeake Bay.

Fortunately there are a number of water current records collected by university researchers in the Chesapeake Bay area, including my institution, the Virginia Institute of Marine Science, Old Dominion University, and the University of Maryland. We will analyze a current record obtained by Old Dominion University scientists at the entrance to Chesapeake Bay (see Fig. 6.1)[6].

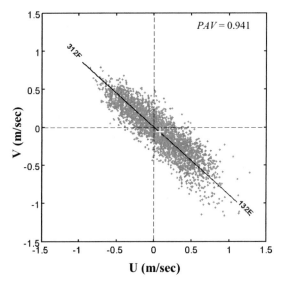

Fig. 7.19. U,V plot of near-surface current, Chesapeake Bay Entrance, North Channel Sept. 2000. White cross marks position of bivariate mean.

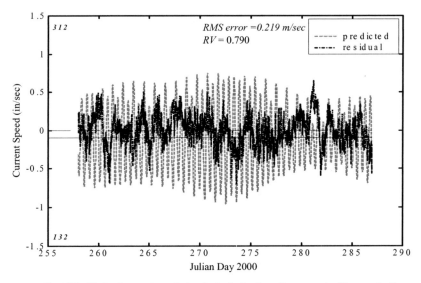

Fig. 7.20. 29-day harmonic analysis of principal axis surface currents, Chesapeake Bay Entrance, North Channel starting September 14, 2000 (JD258). Mean current = -0.10 m/sec in the ebb (132^0) direction.

---

[6] Unpublished data courtesy of Dr. Arnoldo Valle-Levinson, Center for Coastal Physical Oceanography, Old Dominion University, Norfolk, VA.

The surface currents at the Chesapeake Bay Entrance (Fig. 7.19) display more of an elliptical pattern than those seen at Richmond, California and consequently the amount of variance accounted for by the principal axis current is somewhat less: 94.1 percent as opposed to 99.2 percent at Richmond. We may still consider the current one-dimensional and even choose another axis besides the principal one if it suits our purposes. Here is what we can learn using a single axis approach: In Fig. 7.20, we see that the principal axis currents observed at the Chesapeake Bay Entrance in September and October 2000 clearly contain *sub-tidal* oscillations of a relatively high frequency (the peak period of the residual in Fig. 7.20 is 1.7 days). Not an unexpected result since sub-tidal oscillations in water level are more or less the norm during fall-winter months in lower Chesapeake Bay. The fall of 2000 was energetic as usual at sub-tidal frequencies with current variations occurring more in response to local changes in wind speed and direction. Once again there is a reason why we should not expect the model - the *tidal current model* in this case – to be as capable of predicting the **total current** (tidal plus non-tidal) as it might in other regions lacking this sub-tidal wellspring.

What else can we find that tells us how the model is doing? Comparing *RMS* errors for the Chesapeake Bay and San Francisco Bay examples (0.219 m/sec and 0.202 m/sec, respectively) we see that they are similar and both are perhaps a little high. The residual curve for San Francisco Bay (Fig. 7.18), although it contains no evidence of sub-tidal change, nevertheless seems fairly energetic. Is there a reason why the tidal current model for Richmond might under perform a bit? There is and it appears in the residual periodogram in Fig. 7.21.

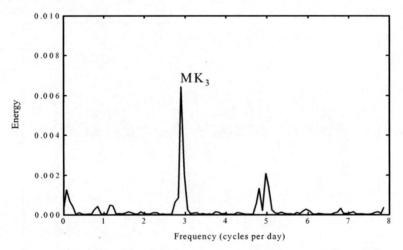

Fig. 7.21. Fourier periodogram of the model residual (observed minus predicted water current), Richmond, San Francisco Bay, November 1, 2002 (JD305).

The compound harmonic MK_3 is a *ter*diurnal constituent (speed 44.025 deg/hr) which we saw a hint of in Fig. 7.7, the periodogram for the tidal residual at Liverpool. It is obviously a significant player in Richmond's observed water currents inside San

Francisco Bay – one that we might want to include in a more comprehensive tidal current model for this station.

Current variations with depth – After our efforts to reduce the dimensionality of horizontal water currents from two to one, it's not exactly welcome news to be told that there may be another dimension to worry about: the vertical position (depth) of current observations and predictions in the water column. Perhaps that's the reason depth is not mentioned in the explanations section preceding daily predictions (Table 1) of the US NOS Tidal Current Tables. Does this mean that we can assume both speed and direction are uniform from surface to bottom? Or (more likely) do the predictions simply apply at an unspecified point near the surface? Actually we do see specific depths given for certain stations listed with current differences and other constants (Table 2) but the first question deserves an answer. To begin with, let's compare the near-surface U,V current for the Chesapeake Bay Entrance (Fig. 7.19) with the near-bottom current at this location during the same time period (Fig. 7.22).

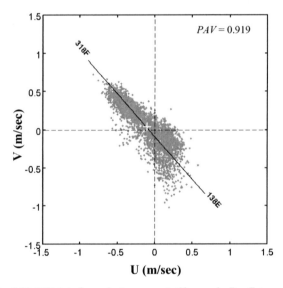

Fig. 7.22. U,V plot of near-bottom current, Chesapeake Bay Entrance, North Channel Sept. 2000. White cross marks position of bivariate mean.

Three things stand out. Firstly, although the principal axis of the near-bottom current in Fig. 7.22 has only a slightly different heading from the one near-surface, we see a pronounced departure from the nearly ellipsoidal current distribution of Fig. 7.19: the ebb component of the near-bottom flow has a 'tail' of current extremes pointed in a much more southerly direction (160°-170° deg.), a sign bottom bathymetry is playing an uneven role in modifying the bottom current. Secondly, the bivariate mean is different, yielding a bi-directional (principal axis) current with net inflow of about 0.10 m/sec near the bottom. Fig. 7.19 shows net outflow of approximately the same amount near the surface consistent with classical two-layer estuarine circulation. Thirdly, the overall magnitude of the current is less near the bottom.

Liverpool Bay

The Coastal Observatory of the UK Proudman Oceanographic Laboratory maintains a set of fixed moorings in Liverpool Bay and the Irish Sea. Their main site outside the River Mersey in Liverpool Bay (Station COA, Fig. 7.23), is an excellent source of water current data. The data are systematically collected and extremely well documented in regular cruise reports now available online at http://cobs.pol.ac.uk/cobs/fixed/. From this site I selected a current record collected at station COA in March 2003. We will examine it at three different depths – or in this case at heights measured in meters above bottom. The water depth reported at station COA is approximately 28 meters.

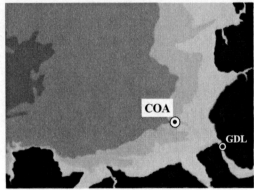

Fig. 7.23. Location of current station COA and tide station GDL (Gladstone Dock Liverpool), West UK.

COA, near-surface (26.7m)
The U,V current plot in Fig. 7.24 has a roughly elliptical distribution but with wider than usual scatter about the mean current (white cross). Unlike San Francisco Bay and the Chesapeake Bay Entrance, this station lies at sea on a shallow shelf, away from the confines of land. Although the principal axis accounts for 87 percent of the total current variance, the ellipse shows signs of 'degenerating' into a circle

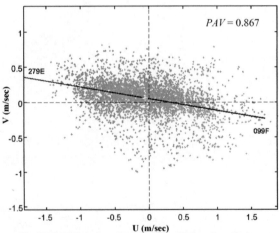

Fig. 7.24. U,V current, station COA, 26.7 m above bottom.

with no principal axis as we have defined it. In that case the tidal current components U and V would divide their total variance equally between any two orthogonal axes. We could simply retain the original east-west and north-south pair and analyze them both. Putting the predicted components together again, we should have a rotating vector - a *rotary current*. However, as we learned in Chapter 6 (Sec. 6.5), the ideal rotary current is expected to change direction continually without ever experiencing a true slack water phase. This concept does not seem at all consistent with the high density of points appearing at *tidal slack water* in Fig. 7.24. Tidal slack water is the 'zero' point we get after removing the tidally averaged current: the white cross in Fig. 7.24. Considering the seeming paradox, it is worth remembering that the non-tidal current we've seen thus far has seldom been a steady current.

Variable winds are particularly active in generating variable currents in near-surface layers. To see the evidence for this type of interpretation we may examine the predicted current and residual curves for the principal axis as displayed in Fig. 7.25. Reduction-in-variance by the tidal current model is low (0.693) and a residual periodogram (not shown) indicates peak energy at a subtidal period of about 9.7 days in addition to a secondary peak at tidal frequency corresponding to the 'Lamda-2' constituent (29.4556 deg/hr). In spite of these explanations for the low reduction-in-variance, this is a case where we might not be justified in treating the current field through a single dimension That said, we should have a look at greater depth at station COA.

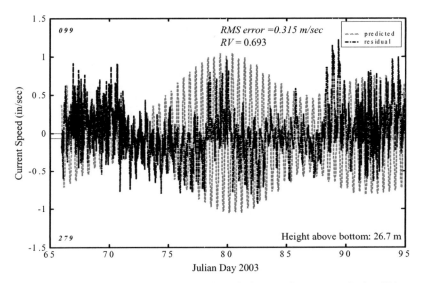

Fig. 7.25. 29-day harmonic analysis of principal axis surface currents, Station COA, Liverpool Bay, Western England starting March 7, 2003 (JD66). Mean current = - 0.07 m/sec in the ebb (279^0) direction.

COA, mid-depth (14.7m)
The U,V principal axis at mid-depth (Fig. 7.26) has rotated counter-clockwise by only 3 degrees relative to the surface axis (Fig. 7.24) but the distribution surrounding it is remarkably different. We're back to a tight ellipse with only minor current variance in any direction other than along the major axis – as though we were in the confines of a narrow river estuary rather than off

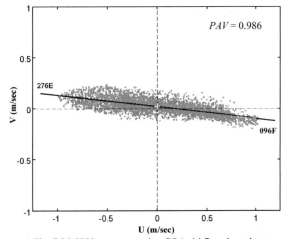

Fig. 7.26. U,V current, station COA, 14.7 m above bottom.

the coast on an open shelf! Wind-induced current is not at all evident in this mid-depth record from station COA but, as at the surface, there seems to be a channeling effect acting toward the coastline at a point just above the nearest estuary, the River Mersey. We also see from an analysis of the principal axis current (Fig. 7.27) that our nine-constituent current model does a good job of accounting for the variance it contains, producing a low *RMS* error as well.

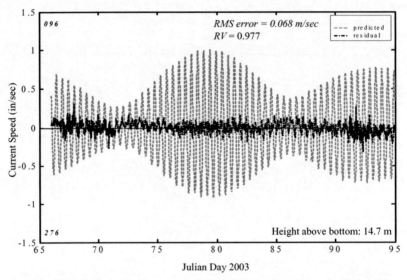

Fig. 7.27. 29-day harmonic analysis of principal axis mid-depth currents, Station COA, Liverpool Bay, Western England starting March 7, 2003 (JD66). Mean current = - 0.01 m/sec in the ebb (276^0) direction.

COA, near-bottom (2.7m)

The *U,V* current shown near the bottom in Fig. 7.28 has a tear-drop shape and an 'eye' near the tidal slack water point where the density of points is low. Evidence of a rotary current? Possibly so. But if there is still a sign of bi-directional current here, it undoubtedly appears in the flows with greatest strength. These are aimed almost due east during what we would definitely call the flood phase.

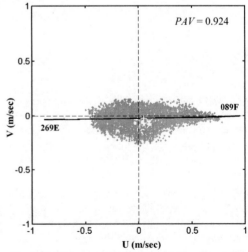

Fig. 7.28. U,V current, station COA, 2.7 m above bottom.

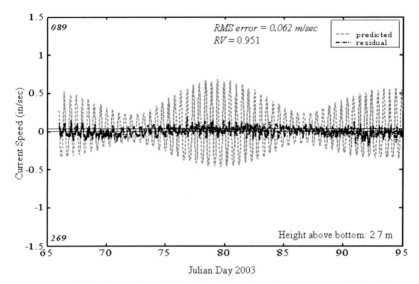

Fig. 7.29. 29-day harmonic analysis of principal axis near-bottom currents, Station COA, Liverpool Bay, Western England starting March 7, 2003 (JD66). Mean current = 0.04 m/sec in the flood (089^0) direction.

Collectively the March 2003 data represent only a single sample of the current up and down the water column at station COA but they tell an interesting story. The near-surface current clearly is unique in this sample because of its strong non-tidal flow components – components not found at mid-depth or near-bottom. If nothing else, we have evidence that the surface current is unlikely to respond as well to conventional prediction techniques regardless of the quality of the astronomical current model. At the same time, the data strongly hint at a consistent astronomical current from surface to bottom. Examining the amplitudes of the nine tidal current constituents used in *SIMPLY CURRENTS* (Table 7.2), there is a consistent decrease in amplitude with depth for most of them. However, their proportions expressed as a fraction of the M_2 amplitude, show little change with depth. Together with the principal axis data, this argues for uniform tidal current characteristics affected only by a frictional attenuation of the current that originates at the bottom.

Table 7.2. Current amplitudes (m/sec), Station COA, Liverpool Bay, Mar 2003. Surface (row 1), mid-depth (row 3) and bottom (row 5). Shaded cells contain ratios of constituent amplitudes to M_2 amplitude.

M_2	S_2	N_2	K_1	O_1	M_4	M_6	S_4	MS_4
0.610	0.248	0.192	0.013	0.024	0.056	0.032	0.010	0.035
1.00	0.41	0.31	0.02	0.04	0.09	0.05	0.02	0.06
0.576	0.263	0.122	0.005	0.014	0.048	0.017	0.005	0.028
1.00	0.46	0.21	0.01	0.02	0.08	0.03	0.01	0.05
0.352	0.163	0.060	0.002	0.010	0.035	0.012	0.008	0.032
1.00	0.46	0.17	0.01	0.03	0.10	0.03	0.02	0.09

7.6 WHEN TIDAL CONSTANTS AREN'T

If you have used *SIMPLY TIDES* or a similar program to analyze the tide one month (29 days) at a time, you may have noticed that a few constituents - the main solar semidiurnal constituent, S_2, being the primary one - appear to undergo a change in both amplitude and phase from one month to the next. At first glance that would seem to be a clear violation of the assumptions of constancy underpinning the astronomical tide model described in Chapter 4 (Sec. 4.6). But there is a reason why this is so and its explanation will carry us over an important hurdle in the path toward a more complete understanding of tidal behavior. The immediate question is: what kind of change are we talking about? We can demonstrate that the change for S_2 is a cyclical one with a period of half a year using a series of water level analyses from the Chesapeake Bay Bridge Tunnel (CBBT) in the year 2002 (Fig. 7.30). Something is causing the S_2 tide wave to act a bit strange but in a systematic way. We will use a simple form of harmonic expansion to find out what it is.

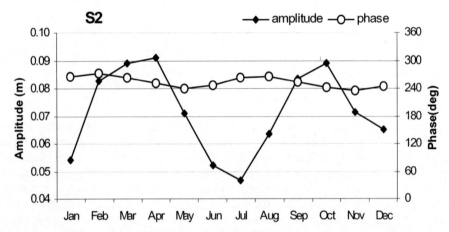

Fig. 7.30. S_2 amplitude and phase determined from harmonic analysis of monthly water level records at the Chesapeake Bay Bridge Tunnel (CBBT) in Virginia during 2002.

Harmonic expansion - The data for the S_2 tidal constituent in the above figure suggest that we are dealing with something other than a simple harmonic. Writing a single term from the astronomical tide model (Eq. 4.2) to represent this constituent, we have

$$h(t) = h_0 + R_1 \cos(\omega_1 t - \phi_1) \tag{7.3}$$

The cosine term on the right becomes a *compound harmonic* if its amplitude R_1, speed ω_1, or phase ϕ_1 - one or more of them - varies with time. In the present instance at CBBT, the speed remains constant but the amplitude and phase each appear to vary periodically about a mean value. Taking just the amplitude, we may represent it with a cycle of amplitude R_2, speed ω_2, and phase ϕ_2 added to its mean amplitude, R_m

$$R_1 = R_m + R_2 \cos(\omega_2 t - \phi_2) \tag{7.4}$$

Substituting Eq. (7.4) in Eq. (7.3),

$$h(t) = h_0 + R_m \cos(\omega_1 t - \phi_1) + R_2 \cos(\omega_1 t - \phi_1)\cos(\omega_2 t - \phi_2) \quad (7.5)$$

The second term on the right in Eq. (7.5) is the one we seek to represent the S_2 tidal constituent with constant tidal amplitude. The third term on the right represents something new that will have to be added to the tide model. Expanding the cosine product in the latter term yields two new harmonic constituents

$$R_2 \cos(\omega_1 t - \phi_1)\cos(\omega_2 t - \phi_2) =$$
$$\tfrac{1}{2} R_2 \cos([\omega_1 + \omega_2]t - \phi_1 - \phi_2) + \tfrac{1}{2} R_2 \cos([\omega_1 - \omega_2]t - \phi_1 + \phi_2) \quad (7.6)$$

The speed of the first new constituent is given by the sum $\omega_1+\omega_2$ and the second by the difference $\omega_1-\omega_2$. We know that the speed for S_2 (ω_1=30.0000 deg./hr) is obtained by dividing 360 degrees by 12.0000 hours, the period of S_2. From Fig. 7.30, we see two complete oscillations in amplitude in one year; hence

$$\omega_2 = \frac{360}{(24 \times 365.24 \times 0.5)} = 0.0821 \text{ deg/hr (solar declination)}$$

The difference $\omega_1-\omega_2$ yields the speed of a very minor constituent of no concern to us here. However, the sum $\omega_1+\omega_2$ = 30.0821 deg./hr is the speed of K_2, an important semidiurnal constituent that derives its origin from the declinational cycles of the moon and (apparent) sun. The component we've just uncovered is related to the sun's apparent motion north and south of the equatorial plane – the cycle that defines our seasons. What about the moon and its declination cycle? A lunar declinational cycle north and south of the equator requires only 27.322 days to complete; thus we have

$$\omega_2 = \frac{360}{(24 \times 27.322 \times 0.5)} = 1.0980 \text{ deg/hr (lunar declination)}$$

If we were to plot monthly amplitude and phase for M_2 just as we did for S_2 in Fig. 7.30, we would see a smaller but nevertheless similar cyclical change. Taking the M_2 speed (ω_1=28.9841 deg/hr), we get $\omega_1+\omega_2$ = 29.9841 + 1.0980 = 30.0821 deg/hr as before. Thus the moon, as well as the sun, contributes to K_2, which is the reason it's called the *luni-solar declinational semidiurnal constituent*.

Again the other harmonic with frequency $\omega_1-\omega_2$ is not important but you may have noticed that Eq. 7.5 seems to contradict this statement by having identical amplitudes, $\tfrac{1}{2}R_2$, for the two harmonic terms on the right. We must recall that ϕ_1 also varies with time, the equivalent of a periodic variation in constituent speed[7]. Variation in speed stems from the complexities of the orbital motions (given cycles in distance as well as declination) and the elements of the tide-producing forces that result from them.

---

[7] The argument ωt is equivalent to $\omega \phi$ with the phase expressed in hours.

Unfortunately, popular concepts of the *equilibrium tide* such as those introduced in Ch. 2 and encountered in most elementary textbooks, shed only a little light on the relationship between orbital motions and tide-producing forces. A more advanced mathematical explanation is required to show that the S_2 tide ought not to vary simply as the angle of apparent solar declination north or south of the equator but as the *square* of the cosine of that angle. That's the reason we see two amplitude oscillations per year rather than one in Fig.7.30. When combined with harmonic expansion, equilibrium tide theory points to a host of additional constituents including longer-period monthly and semimonthly constituents. We will not explore their origins any further in this book.

Perturbations – When two cosine waves are encountered that have nearly the same frequency (speed) but different amplitudes, the one with the smaller amplitude is said to affect the other by causing a **perturbation** in it whose severity depends on the difference in speed. Thus the perturbation of S_2 that appears in Fig. 7.30 is relatively large because the disturbing constituent K_2 has a speed that differs by a mere 0.0821 deg/hr. From the analysis standpoint there are now two additional problems: we not only have another constituent to include in our astronomical tide model, we also have to analyze a longer record to determine it and to effectively remove the perturbations it causes on other constituents. Fig. 7.30 shows that we need a minimum of six months to get a firm estimate of the mean amplitude for S_2. Similar reasoning tells us that we will require a continuous record of at least six months duration to get a reasonable estimate of amplitude and phase for K_2 through least squares harmonic analysis.

What if adequate records aren't there? A continuous, six-months record of water current can be especially difficult to obtain. In the past, so-called inference methods were used to infer constituent amplitude and phase values for a number of diurnal and semidiurnal constituents. In these methods, it was assumed that an equilibrium tide model could approximate amplitude ratios and phase relationships between constituents of the same type, even though the theoretical values might differ widely from those actually observed at a given place. For example, an older manual of harmonic analysis still used in the US gives 0.272 as the expected amplitude ratio between K_2 and S_2. This ratio allows K_2 amplitudes to be inferred from those of S_2 but the problem that remains is – how do we first eliminate the perturbations affecting S_2? There are formulae for elimination, as it is called, but vector averaging of constituents over a four to six-month period may be a better alternative.

After much work to remove the perturbations and determine mean values for their tidal constants, the combination of S_2 and K_2 in a predictive model puts them back again! We scarcely notice it among the more prominent cycles in the predicted tide but, thanks to M_2 and K_2 interaction, there is a variation featuring tides of slightly greater range during the spring and fall equinoxes coupled with tides of lesser range during the summer and winter solstices. You can see that as well in Fig. 7.30.

One final note concerning perturbations: recall the 18.6-year cycle of the lunar nodes discussed in Chapter 2? It's another example of periodicity in lunar declination. We could again use harmonic expansion but here the period is so long we scarcely notice the resultant perturbations on lunar constituents within any given year. Instead it is customary to apply a nodal factor and phase adjustment to convert the 18.6-year mean amplitude and phase of the lunar constituents to year-specific values and vice-versa. The details of this conversion are presented in Chapter 8 (Sec. 8.1).

8

Predicting tides and currents

Tide and current predictions fall into two categories: those that are official and those that aren't. In the United States, NOAA and the National Ocean Service (NOS) have sole responsibility for making official predictions. In the United Kingdom, the Admiralty Tide Tables are published under the authority of the UK Hydrographic Office. This is not to say that their tide and current predictions are necessarily better than anyone else's, but it reflects the fact that they are on the hook when it comes to matters of marine safety and navigation. You and I would rather not be the author of exhibit A in a maritime accident hearing. Consequently, nothing in this book is intended to replace the official information sources customarily used in vessel navigation.

That said the art of tide and current prediction is open to anyone with an interest in the marine environment, from sailors and scientists to saltwater fishermen and waterfront property owners. Advancements have been made in recent years that go well beyond printed tide tables that still require the user to add, subtract, and multiply time and height corrections by hand, just to get the daily highs and lows. New alternatives *Tides Online* (http://tidesonline.nos.noaa.gov), *Easy Tide* (http://www.hydro.gov.uk) and other commercial software products available from the private sector use computer graphics of the type presented in Chapter 7 to display predicted tidal heights and current speeds in a time series format. Moreover, these predictions are made using official harmonic constants such as those found on the NOS CO-OPS web site described in Chapter 7 (http://co-ops.nos.noaa.gov). So, why not use the official tidal constants[1] ourselves to make tidal predictions? One reason why not is that it's a more complicated task than you might think, thanks in part to older methods dating long before the age of digital electronic computing. The following section explains why this is so.

8.1 THE US NOS TIDAL PREDICTION FORMULA

NOS tidal height predictions are based on a harmonic equation with several arguments. For a single tidal constituent at a selected location, the predicted tidal height above mean sea level at time t is

$$h(t) = fH \cos(\omega t + V_0 + u - \kappa) \qquad (8.1)$$

where H is the *mean* constituent amplitude, f is a factor that reduces the mean amplitude to its value in a given year, $V_0 + u$ is the constituent's *equilibrium* phase, and κ (kappa) is the constituent phase lag or *epoch* expressed in degrees. The speed of the constituent, ω (omega), is given in degrees per mean solar hour and t is serial time in hours from midnight starting the year of predictions. Although Greenwich Mean Time (Universal

---

[1] Official harmonic constants are not available to the public in the United Kingdom.

Standard Time) is most commonly used for vessel navigation at sea, Local Standard Time (LST) is generally the preferred choice in coastal areas.

In Eq. 8.1, H and κ are constants but the **nodal factors** f and u vary from year to year for the lunar constituents as stated in Chapter 7 (Sec. 7.6). Variation in f reflects the fact that lunar constituent amplitudes vary periodically about a mean value during the regression of the lunar nodes through their 18.6-year cycle (see Chapter 2, Sec. 2.8, Fig. 2.12). It therefore varies from year to year in accordance with certain celestial formulas (e.g., f varies between 1.038 and 0.963 for the major lunar semidiurnal constituents). The lunar equilibrium phase locates the position of the moon with respect to a given meridian (usually the Greenwich meridian). Two symbols are used to indicate that the lunar phase has a uniform component V_0 that changes at the speed of the constituent as well as a variable component u that adds a small periodic variation (approximately ±2° for the major semidiurnal constituents) during the 18.6-year cycle. The zero subscript in V_0 indicates that the symbol refers to the constituent phase at the particular origin of time ($t = 0$) in use in Eq. 8.1. Let's explore what this means.

The equilibrium tide, as explained in Chapter 2, is a fictitious tide – one that calls for the high water phase to occur just as the body producing the tide (sun or moon) passes the observer's local meridian, or its counterpart on the opposite side of the earth. For example, the M_2 equilibrium tide, represented by the dashed line in Fig. 8.1, reaches its high just as the moon passes overhead. The phase $V_0 + u$ shown in hours[2] is marked by the high water position relative to the time origin ($t = 0$). The constant κ then appears as the lag between the equilibrium phase and the arrival of the next (actual) high water at the observer's location. The phase ϕ that results directly from harmonic analysis (Eq. 4.2) is then found as the difference $\phi = \kappa - (V_0 + u)$.

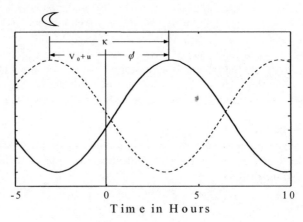

Fig. 8.1. Comparison of phase for the equilibrium tide (dashed line) and the observed tide (solid line) at $t = 0$.

Calculating $V_0 + u$ and the nodal factor f for a set of tidal constituents is done with certain well-known formulas from celestial mechanics. These bear the name *orbital elements*. The ones required for nodal corrections in tidal analysis and prediction are presented in Appendix 2 of this book. The calculations are an added task, of course, and

---

[2] Phase in hours equals phase in degrees divided by constituent speed.

the results differ from one location – one local meridian - to the next. In the age before high speed computers, it made sense to compute orbital elements for a span of years and to publish tables of $V_0 + u$ values for the Greenwich meridian only - in place of the observer's local meridian. The result is called the **Greenwich equilibrium phase**. A modified epoch, written as κ', can then be used in place of κ in Eq. 8.1, enabling predictions using Greenwich $V_0 + u$. The modified epoch is computed as

$$\kappa' = \kappa + jL - \omega S$$

where j is the tidal species number (1 for diurnal constituents, 2 for semidiurnal, etc.), L is the longitude of the local meridian, ω is the constituent speed, and S is the time difference in hours between GMT and local standard time at the place in question.

Today there is another choice – an alternative to the Greenwich equilibrium phase. Rather than choosing a standard meridian, it is possible to choose a standard *time origin* for tidal predictions. As we will see in the next section, there is an advantage inherent in many software packages (e.g., spreadsheet programs) that make use of *serial time* and a single time origin (e.g., midnight at the start of the twentieth century) to keep track of date and time.

8.2 HARMONIC CONSTANTS WITHOUT EQUILIBRIUM ARGUMENTS

The set of equilibrium arguments described above are needed in their entirety only if we plan to change the time origin to midnight of each year in which predictions are made. But if we keep the same time origin from one year to the next, V_0 becomes, in effect, a constant and may be combined with κ in Eq. 8.1 to obtain a new constant

$$\kappa^* = \kappa - V_0 \tag{8.2}$$

or, since $\phi = \kappa - (V_0 + u)$,

$$\kappa^* = \phi + u \tag{8.3}$$

The new harmonic constant, κ^*, is derived after calculating u from celestial formulas governing nodal variations at the mid-point of the year of observations (see Appendix 2). When making tidal predictions for a different year, u must be calculated anew so that ϕ in Eq. 4.2 is derived as

$$\phi = \kappa^* - u$$

The modified NOS tide prediction formula then becomes

$$h(t) = fH \cos(\omega t + u - \kappa^*) \tag{8.4}$$

Since we are adding u in one year and removing it in another, the reference meridian is immaterial so long as it is not changed in the process; once again, f and u account for nothing more than a small and slowly varying amplitude and phase change, respectively, that occurs among the lunar constituents during the 18.6-year nodal cycle. One must be aware of the difference in the modified formula (Eq. 8.4) only when attempting to compare κ^* with κ' at a particular NOS tide station. Obviously NOS phase arguments cannot be used unmodified in programs based on Eq. 8.4 but two rather special tidal constituents from the list of NOS harmonic constants available for local tide stations can be used 'as is' in my tidal analysis and prediction program *Simply Tides* (Sec. 8.3). These are the *solar annual constituent*, Sa, and the *solar semi-annual constituent*, Ssa, whose periods are the mean solar year and one-half the mean solar year, respectively. While these constituents have an astronomical origin, they are augmented by seasonal variations in water level that arise from steric (water density) effects. Steric effects add to the astronomical constituent amplitudes for Sa and Ssa because they exhibit the same annual and semi-annual cycles of variation. And the additions, in most instances, are not small.

Harmonic constants for Sa and Ssa require a substantial amount of data to determine. They are usually estimated by analyzing a special time series consisting of twelve consecutive monthly mean sea level values (January through December), each monthly mean representing an arithmetic average for that month over a number of years[3]. Values of the Greenwich equilibrium phase $V_0 + u$ for Sa and Ssa in the year 2000 were $280°$ and $200°$, respectively; these values change only slightly (by less than a degree) in several years time and their nodal factor, f, is always unity. So after finding their amplitude H and epoch κ' in the NOS tidal constants list, the seasonal tides Sa and Ssa are easily included in the predictions as explained in the next section. But remember that the tidal constants for Sa and Ssa are *estimates* based on oceanographic conditions averaged over several years. Even if the long-term average is consistent over time, Sa and Ssa may provide only a rough estimate of the actual seasonal tide experienced in any given year.

8.3 TIDAL HEIGHT PREDICTIONS MADE SIMPLE: *SIMPLY TIDES*

A suite of my tidal prediction programs may be accessed through a single Graphical User Interface (GUI) called *SIMPLY TIDES* written in the MATLAB® programming language. The programs perform both analysis and prediction of tides based on Eq. 8.4. A detailed description of the GUI, and its method of use, appears in Appendix 3 (a command line version appears in Appendix 5). MATLAB graphics of the type shown throughout this book are an integral part of *SIMPLY TIDES*. For example, Fig. 8.2 contains a plot of predicted hourly tides at the Chesapeake Bay Bridge Tunnel (station CBBT) for the year 2003. Although we can't see the daily information it includes at this scale, the graph clearly illustrates the interaction of the spring-neap and apogean-perigean cycles and their fortnightly effect on tidal range during the year.

While the seasonal water level change (sum of the constituents Sa and Ssa), is included in the predictions (gray curves), it is drawn separately as the black curve in Fig. 8.2 to underscore its contribution to the total water level on a scale of months. Mean water level is expected to vary by about 19 cm (0.6 feet) between a seasonal low on Julian day 18 (January 18) and a seasonal high on Julian day 271 (September 28). Unless and until NOS revises the harmonic constants for Sa and Ssa, the same seasonal pattern of water level variation will be predicted every year at station CBBT.

---

[3] The monthly means are calculated as deviations from the annual mean water level in each year; Fourteen years of hourly heights were used to determine Sa and Ssa at the Chesapeake Bay Bridge Tunnel (CBBT) tide station.

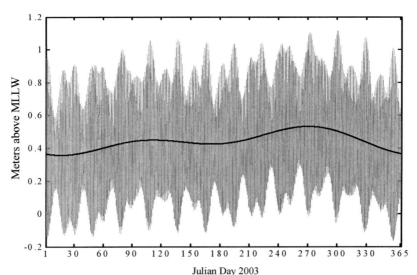

Fig. 8.2. Predicted tides at Chesapeake Bay Bridge Tunnel in Virginia for the year 2003. The black curve is the seasonal tide (Sa+Ssa).

8.4 COMPARISON OF *SIMPLY TIDES* WITH US NOS TIDE PREDICTIONS

It's fair to ask how well the two methods compare in terms of the predictions they make. NOS predictions normally employ harmonic constants for up to thirty-seven tidal constituents, the number obtained in a standard NOS 369-day analysis. Although the *SIMPLY TIDES* program can be easily modified to include additional constituents, the present nine are considered optimal for a 29-day analysis without recourse to inference methods. Starting with the assumption that thirty-seven constituents will predict tides more accurately than eleven (including Sa and Ssa), the question becomes: "what do we lose by using only eleven?" Using a sample of analyzed tide data as the basis for comparison, the answer may surprise you.

Once the harmonic constants are accepted, tide predictions by NOS formulas - as well as those by *SIMPLY TIDES* - are fully deterministic; it isn't necessary to compare simultaneous predictions at great length to find an answer to the above question. A four-day comparison is perhaps adequate for our purposes using the data that appear in Fig. 8.3. Predictions by *SIMPLY TIDES* (black curve in Fig. 8.3) were generated using harmonic constants obtained in August, 2002, in conjunction with tidal datum elevations supplied by NOS. The NOS harmonic constants for CBBT represent the average of five 369-day analyses between 1976 and 1989. NOS predicted times and heights of high and low tide are shown as gray ovals in Fig. 8.3. The shape of the ovals marking the NOS highs and lows conveys the approximate limits of accuracy at this station. The predictive accuracy of high and low water times for Hampton Roads stations is reported in NOS Tide Tables to be 0.4 hours at the 90% distribution level. Heights for comparative purposes are limited by quantization since they are officially reported only to the nearest 3 cm (0.1 feet).

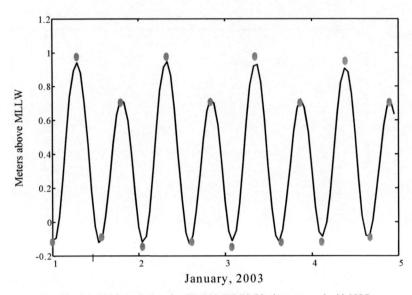

Fig. 8.3. Tidal predictions by *SIMPLY TIDES* (black) compared with NOS predicted times and heights of highs and lows (gray ovals) at CBBT.

The phase (time) agreement between the two methods appears quite good and a high water diurnal inequality is also well represented by both methods. Some discrepancy is noted, however, in the extremes. *SIMPLY TIDES* appears to under predict most of the low waters by approximately 2 to 3 cm and to under predict the higher high waters by approximately 4 to 5 cm. Although the agreement is not bad, is this is the best a 29-day analysis can do? Not at all. Tidal predictions made with *SIMPLY TIDES* can be further improved by vector-averaging the harmonic constants from several consecutive 29-day analyses. To do this, amplitude and phase vectors expressed in polar coordinates must first be converted to Cartesian coordinates:

$$X = R\cos\phi; \quad Y = R\sin\phi.$$

Vector averaging requires the mean of X and Y to be determined for each constituent. We can determine their combined variance as well and use the latter to calculate standard deviation and standard error, S_e, before converting the result back into polar coordinates as vector-averaged R and ϕ. However, one other bit of information is needed for a fair comparison.

NOS buildup factors – The NOS considers that some of the variance unaccounted for in a least squares harmonic analysis should be made up by applying '*buildup factors*' to the tidal constituent amplitudes that result from the analysis. They therefore multiply all of the constituent amplitudes, except those for Sa and Ssa, by a constant before using them in tidal predictions. At CBBT, the factor used for this purpose is 1.04. The justification for introducing this particular bias may rest with risk avoidance concerns. In a legal context, there is risk in under-prediction of high and low tides; e.g., a ship

Sec. 8.4]   **Comparison of *SIMPLY TIDES* with US NOS tide predictions**   129

may run aground because – it is claimed – the tide was lower than predicted. But in the present scientific context it is in our interest to remove this bias by dividing the original NOS constituent amplitudes by 1.04 before making a final comparison. Table 8.1 lists the corrected NOS amplitude means together with the corresponding *SIMPLY TIDES* amplitude means and confidence intervals. The data in this table allow us to test a null hypothesis: the hypothesis that the mean difference for each constituent is *zero*.

Table 8.1. Comparison of NOS, *SIMPLY TIDES* amplitude means, mean difference and confidence intervals ($\alpha=0.05$) in meters for nine constituents at CBBT.

SOURCE	M2	S2	N2	K1	O1	M4	M6	S4	MS4
NOS	0.389	0.071	0.088	0.053	0.040	0.006	0.007	0.003	0.005
SIMPLY	0.393	0.062	0.087	0.053	0.037	0.006	0.007	0.002	0.003
Diff.	-0.004	0.009	0.001	0.000	0.003	0.000	0.000	0.001	0.002
$t_{.05}S_e$	0.010	0.020	0.020	0.015	0.006	0.002	0.001	0.003	0.003

Accepting the NOS 369-day harmonic analysis as the standard for comparison, we will assume the corrected NOS amplitudes in the top row of Table 8.2 are without error. Subtracting the second row containing *SIMPLY TIDES* vector-averaged amplitudes from the first row of NOS amplitudes yields the differences shown in the third row. The confidence intervals in the fourth row were computed as the standard deviation of the mean, S_e, times the t statistic for a two-tailed test at the 95 percent confidence level ($\alpha=0.05$). In every column the mean difference is less than the confidence interval. In other words, the null hypothesis that the NOS and *SIMPLY TIDES* amplitudes are the same cannot be rejected on the basis of the present data. These data indicate that the means are not significantly different.

Tests with actual tides - The real test of either method, of course, lies in how well they predict actual tides. The NOS procedure nominally can be expected to do a better job, given the greater power of a vector-averaged 369-day analysis and its extra tidal constituents, even though the predictive differences in Fig. 8.3 appear to be due to the buildup factors used by NOS.

With these differences in mind, it's worth examining *SIMPLY TIDES* predictions from Fig. 8.3 compared to the actual tides at CBBT during the same four day period, as shown in Fig. 8.4. Water levels at CBBT during the first four days of January 2003 were strongly influenced by a weather event. Winds were calm at the beginning of January 1 but gusts quickly increased to 11 knots from the southeast, further increasing to 20 knots from south-southeast by mid-morning, accompanied by an atmospheric pressure drop from 1016 mb to 1000 mb by mid afternoon. On January 2, winds remained above 15 knots and continued to change direction from west around to the northeast – the classical development pattern of a 'northeaster'. The dominance of the winter storm event is evidenced by the smooth rise and general lack of 'chaotic' water level fluctuations in Fig. 8.4, a fact that makes it useful for comparison with the predicted tide.

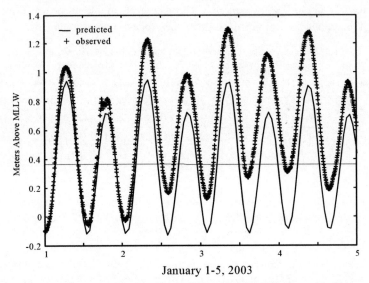

Fig. 8.4. Tidal predictions by *SIMPLY TIDES* (solid) compared with observed 6-minute water levels (crosses) at CBBT.

The first point to be taken from Fig. 8.4 is that the *phase* of the observed and predicted semidiurnal tides are in good agreement; the times of the highs and lows appear well matched throughout the four-day period. The second point is that water level heights are fairly consistent where we would expect them to be. Clearly the *observed* water levels are raised overall, being the same as predicted water levels only near the beginning of January 1. Otherwise, individual *range* values (high water – low water differences) diverge in some instances but agree well in others and the diurnal inequality is much the same for the predicted and observed tides. The reality seems to be that we can expect only modest gains through further refinement of the astronomical tide model as compared to a much larger gain if we could only come up with an effective model of the meteorological tide! That is possible through hydrodynamic modeling but such a task lies beyond the scope of this book.

Secular sea level change - One other factor requires clarification when comparing observed tides with predicted ones: long-term (secular) sea level change. In addition to an adequate model of the meteorological tide, it would be useful to have a reliable way of monitoring secular trends in sea level rise or fall, a subject introduced in connection with tidal datum definition in Chapter 5. As stated in that chapter (Sec. 5.7), the most recent linear trend at Hampton Roads, Virginia from 1927 through 1999 suggests that mean sea level is now rising at a rate of 4.25 mm/year (1.39 feet/century) relative to the land. Thus in the year 2003, thirty-four years from the mid-point of the 1960-1978 Tidal Datum Epoch, mean sea level in the lower Chesapeake Bay should have risen by about 15 cm (6 inches). As one would expect, observed water levels that reference height relative to the land, or to a fixed datum that does *not* keep up with sea level rise, will increase with time. The problem becomes quite apparent when you compare recent observations referencing an older datum (e.g., 1960-1978 *MLLW*) with tidal predictions. The reason is simple: most tidal predictions take no account of secular sea level trends on a year-to-year basis as they do for tidal range variation during the 18.6-year nodal cycle. In the tidal prediction model, *MLLW* is simply an offset from *MSL*; neither *MLLW* nor *MSL* have any intrinsic connection to the land in the context of the model. As it happens, 1960-1978 *MLLW* was still in effect when I obtained the data for Fig. 8.4 and it was necessary to subtract 15 cm from each of the NOS water level observations reported at CBBT before plotting them in that figure. Real-time comparisons of predicted versus observed tides, comparisons such as those appearing on the NOS PORTS and *Tides Online* web pages, make a similar correction.

8.5 TIDAL CURRENT PREDICTIONS MADE SIMPLE: *SIMPLY CURRENTS*

Currents can be predicted with a companion program to *SIMPLY TIDES* named *SIMPLY CURRENTS*. Program *SIMPLY CURRENTS* analyzes either a 29-day or a 14-day current record and automatically stores the resulting harmonic constants for the *principal axis currents,* the rotated U_p component of the water velocity field. Methods for deriving the principal axis current are explained in Chapter 7. The *mean current* in the direction of the principal axis during a 14-day analysis is usually not a representative measure. It should normally be reset to zero by the user – unless there are other measurements available that adequately represent the non-tidal current. For example, in the uppermost reaches of a river estuary, current speed may fluctuate at tidal periods without a reversal of current direction. The non-tidal current in this instance is represented by river inflow that exceeds the tidal current.

A command line version of *Simply Currents* appears in Appendix 6.

8.6 COMPARISON OF *SIMPLY CURRENTS* PREDICTIONS

Observations available from NOS current stations such as Richmond, CA, in San Francisco Bay (designated station RISB in Chapter 7) may be used for comparison with *SIMPLY CURRENTS* predictions. Of course, an independent series of observations should be selected apart from the 14-day series starting November 2, 2002, (Fig. 7.12) used to obtain the harmonic constants at station RISB. Accordingly, a short, three-day series beginning January 1, 2003, was selected for comparison with *SIMPLY CURRENTS* predictions for this period.

Allowing for the relatively high noise level associated with the 6-minute current measurements, *SIMPLY CURRENTS* (using nine harmonic constants) generates predictions that agree very well with the observed currents shown in Fig. 8.5. The usual NOS current predictions for station RISB during this period include slack water times and the times and speeds of maximum flood and ebb. These predictions also agree with the observations; e.g., NOS predictions for the three maximum ebb currents are -2.8, -2.8, and –2.7 knots, respectively, in the order of their occurrence in Fig. 8.5.

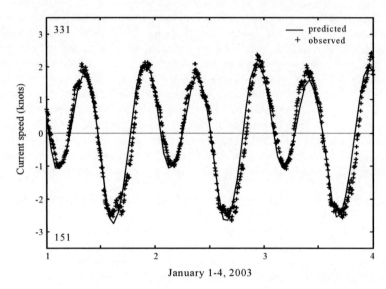

Fig. 8.5. Tidal current predictions, *SIMPLY CURRENTS* (solid) compared with observed 6-minute currents (crosses) at station RISB.

It is apparent, however, that among the 6-minute observations shown in Fig. 8.5, those appearing at or near the points of maximum current display the greatest amount of noise and outright deviation from the smooth sinusoid that is the basis of harmonic predictions. Is this due to some quirk of the instrumentation? Or is it real? Regardless of the answer, it makes the job of comparison particularly difficult at those points. Slack water times appear well defined in Fig. 8.5 but comparing them to NOS reported times involves some uncertainty if the flood and ebb headings are not exact reciprocals of one another. Not that we shouldn't make these comparisons, but to properly assess the fit of the astronomical model to the observations, we need to look at the rest of the curve as well as the extremes.

8.7 THE OUTLOOK FOR MORE AND BETTER CURRENT DATA

Observed current data are difficult to find in most of the coastal estuaries of the United States at present. This is likely to change as coastal monitoring programs expand and the technology improves. It is difficult to say when this will occur because the cost of new current measuring equipment, e.g., the *acoustic Doppler profiler* (ADP), is quite high, as is the cost of deploying and servicing these instruments, to say nothing of the subsequent task of data quality control and verification prior to its release.

There are some convincing reasons why the need for long-term water current measurements of high quality will expand in the future. In addition to ongoing port and harbor development plans, issues addressing homeland security in the nation's coastal waterways are moving to the forefront of attention. This means that the need for accurate tide and current information will no longer be restricted to vessel navigation and traffic management issues. Concern is now growing rapidly over the fate of hazardous materials introduced into our waterways; materials introduced not only by the usual industrial and commercial sources. There are now some new and potentially catastrophic ones amid the rising threat of global terrorism. Knowing where water masses and water-borne materials are going, and how quickly the material is being dispersed, is no longer merely an academic matter.

Current measurements made at a single point are insufficient to address problems of this kind by themselves. The necessary coverage of space and time in estuarine systems can only be achieved with three-dimensional hydrodynamic models. To provide these models with the capability to make accurate predictions over relatively short periods of future time - so-called **nowcasts** - the technique of **data assimilation** is required. Nowcasts are 'rolling' predictions continually revised and updated with real-time data from the field, data that include current speed and direction measured at fixed locations. The hydrodynamic model in turn produces a huge field of current predictions filling in the time and space of the model domain – a gigantic multiplier of the data gathering effort.

9

Storm tide and storm surge

9.1 BEYOND THE ASTRONOMICAL TIDE

Tides are not the only contributor to changing water levels in the sea. Over the short-term (hours, days, weeks), just about everything that isn't part and parcel of the *astronomical tide*, the tide caused by the moon and sun, seems to play a role in shaping the *meteorological tide*, the change in water level that can be linked to changes in the atmosphere – the weather and its numerous effects. An exception is the *tsunami*, the so-called tidal wave unleashed by undersea earthquakes and volcanic eruptions. But like a tsunami, an unusual weather event can trigger an extreme change in water level at the coastline as well. When that happens, a specific term applies: storm surge. A life-threatening storm surge is most often associated with a major **hurricane** (Atlantic ocean), **typhoon** (Pacific ocean), or **tropical storm**. However, another type of storm, the **extratropical cyclone** (*northeaster* in eastern US, extratropical storm in western Europe) can produce damaging surge both high in elevation and sprawling over large areas, often including several US coastal states in a single storm. Unlike tropical storms that originate as easterly waves traveling west over the tropical ocean, extratropical storms in North America usually begin life as a low-pressure frontal wave between two air masses over the continent that subsequently moves in a northeasterly direction. The foremost example of a severe extratropical cyclone in the US is the *Ash Wednesday storm of 1962*, an unusually slow-moving or 'blocked' weather system that battered the east coast from North Carolina to Long Island, New York for more than two days.

Hurricanes, although more intense, tend to concentrate their fury around the 'eye' of the storm, as did hurricane *Andrew,* whose eye can be clearly seen in the NOAA satellite image at right, and wallop the unfortunate localities that happen to lie in front of them as they make their landfall. US Weather forecasters rely on storm surge models running in real time to predict storm landfall sites as well as the expected intensity of the storm and its effects at those sites. In recent years these models have provided very timely warnings to coastal communities most at risk using a number scale of one to five.

Fig. 9.1. NOAA satellite image of hurricane Andrew, August 23, 1992.

134

9.2 THE US SAFFIR-SIMPSON HURRICANE SCALE

Thanks to US television news, NOAA's National Hurricane Center has become well known to just about every coastal resident of the eastern United States. After long experience in tracking hurricanes, forecasting their strength and predicting the risks they present, the Center devised a 1-5 rating scheme known as the **Saffir-Simpson Hurricane Scale**. Although the five categories that make up the scale are based on wind speed, each has a carefully described set of characteristics, including the height of the storm surge that typically occurs (Table 9.1).

Table 9.1. Saffir-Simpson Hurricane Scale, Wind Speed and Storm Surge Limits.

	CATEGORY 1	CATEGORY 2	CATEGORY 3	CATEGORY 4	CATEGORY 5
WIND	74-95 mph 64-82 knots	96-110 mph 83-95 knots	111-130 mph 96-113 knots	131-155 mph 114-135 knots	>155 mph >135 knots
SURGE	4-5 feet 1.2-1.5 m	6-8 feet 1.8-2.4 m	9-12 feet 2.7-3.7 m	13-18 feet 4.0-5.5 m	>18 feet 5.5 m

Experience shows that storm surge heights are by no means uniform over the landfall area. For example, hurricane *Andrew*, a Category 5 hurricane with winds greater than 155 mph, struck south Florida with devastating force on August 24, 1992, and produced a 5.2 m (17 ft) storm surge near its landfall site just south of Miami. Barely 100 miles away in Key West, Florida, the recorded storm surge was hardly noticeable.

9.3 SOME DEFINITIONS

Storm surge, by definition, is the change in water level that would occur in the absence of the astronomical tide during a storm's passage through a given coastal area. It can be positive or negative relative to the mean water level that one chooses to define, and there is no minimum in terms of its magnitude. Of course, tides don't take holidays and the storm tide (the highest water level that results during the storm) is a combination of both the storm surge and the astronomical tide that simultaneously occurs. We can make an estimate of the storm surge history (its height through time during the storm, sometimes called a *surge hydrograph*) by simply subtracting the predicted tide from the observed tide at a local tide station. This is not as easy as it sounds for several reasons. For one thing, hurricanes have an excellent record of destroying tide gauges that get in their way. For another, the storm surge may interfere with the astronomical tide, causing it to change slightly from the tide that would have resulted in the absence of the storm. Finally, as was the case for hurricane *Andrew*, the height of the storm surge can vary tremendously from point to point within a relatively small area. Since there are not that many continuously operating tide stations even in the United States, the odds are against any one station capturing the absolute maximum storm tide in the region.

For the above reasons, maximum storm tide heights are usually determined from high water marks left inside buildings that survive the storm. This is where things get tricky. Unlike a tide gauge that has a stilling well or a similar device for filtering out waves, high water marks include not only the storm tide (the astronomical tide plus the storm surge) but may also include an additional contribution from wind waves. No one really knows for certain what these waves may be like in the center of a major hurricane making a landfall. We see only the evidence they leave behind in flooded buildings that often display variable high water marks because some interior sections behaved more like a stilling well than others during the storm: as a result, heights have been known to vary by as much as 15 cm within a single building. But even the lower marks found lead to some distressingly high numbers for the storm surge heights filed in official reports.

9.4 STORMS OF RECORD IN THE UNITED STATES

In spite of the difficulties, it's tempting to look inside the box and see what the tide record has to say about water level behavior during a major storm event. Records from the Ash Wednesday storm provide a good look at an extratropical event and other records are available to illustrate tropical storm and hurricane events. I know of a good example of the latter type from personal experience.

Hurricane *Camille*: This is a storm I'm unlikely to forget. In June of 1969, I moved my family out of our seaside apartment in Pass Christian, Mississippi, to go to a new job in Virginia. I was leaving my old job as a NOAA officer in charge of a hydrographic field party charged with conducting nautical chart surveys in Mississippi Sound. Two months later *Camille*, a Category 5 hurricane, crossed the Sound and flattened most of Pass Christian, including our old apartment building. Nothing was left of it but a bare concrete slab with broken pipes standing like reeds in a marsh. In addition to the terrible damage done by this storm, the change to the bottom topography of Mississippi Sound was profound, rendering much of our survey work obsolete.

According to NOAA records, hurricane *Camille* produced a storm tide reaching 7.5 m (24.6 ft) above *MLLW* at Pass Christian. No tide gauge was left intact anywhere along the Mississippi coast so leveling surveys measuring several high water marks left after the storm provide the only evidence of this tremendous storm tide. Some of the tidal benchmarks in Pass Christian apparently survived but with questions about their stability. It's likely that their datums were later re-established with new water level data in order to recover *MLLW* for referencing the height of the storm tide marks found on the insides of the remaining buildings.

Besides *Camille* and *Andrew*, only one other Category 5 hurricane, the *Florida Keys 1935* hurricane, appears on the record books among storms striking the continental United States. So it's not at all surprising that the nearest surviving tide record that shows something of the true effect of hurricane *Camille* comes from Pensacola Bay, Florida, about 125 miles east of Pass Christian. Even at that distance, the storm tide recorded at Pensacola is an impressive one with a sharp, pulse-like surge as Fig. 9.2 clearly shows[1].

---

[1] Fig. 9.2 also shows a small but persistent diurnal oscillation in the residual curve, an indication that certain additional diurnal constituents should be combined with K_1 and O_1 in the tidal prediction model used at Pensacola. The principal features of the storm surge are evident nevertheless, particularly those arising during equatorial tides.

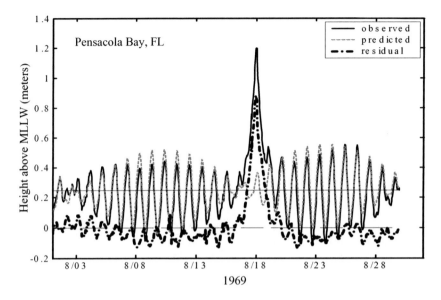

Fig. 9.2. Storm tide from hurricane *Camille* recorded at Pensacola Bay, Florida.

At Pensacola Bay, *Camille*'s storm tide briefly attained a height of about 1.2 m (3.9 ft) above *MLLW* as shown by the sharp peak in Fig. 9.2. That's quite a bit less than the 7.5 m storm tide at Pass Christian but still well above the normal range of the diurnal tide at Pensacola Bay. And it could have been worse. Because the tide at Pensacola Bay is almost completely *diurnal* (one high tide and one low tide per lunar day), the minimum range during *equatorial tides* (when the moon is on or near the equator) is quite small. By chance, hurricane *Camille* produced the maximum storm surge at Pensacola Bay during an equatorial tide. Had it occurred earlier, say on August 10, or later on August 25, the storm tide would have been about 30 cm - nearly a foot - higher. On a low-lying coast with shallow-sloping ground, even a relatively small rise like this can move the waterline (and the wave action that accompanies it) farther inland.

The Ash Wednesday storm – This storm began as a coalescing low-pressure system in the southeastern part of the United States. The deepening low traveled northward and was met by a high-pressure system that halted its progress just offshore in the Middle Atlantic Bight where strong cyclonic winds blew uninterrupted for almost three days starting March 7, 1962 ('Ash Wednesday' marking the beginning of Lent). While it is not uncommon for a *northeaster* to produce winds approaching hurricane force, it is very unusual to encounter one that produces them for more than one day. In the case of the 1962 storm, the extra duration more than made up for sub-hurricane strength winds and generated a hurricane-class storm tide aided by astronomical spring tides (Figs. 9.3 and 9.4). At Hampton Roads, Virginia (Fig. 9.3), the maximum storm tide reached 2.2 m (7.2 ft) above *MLLW*; at Lewes, Delaware (Fig. 9.4) near the Atlantic entrance to Delaware Bay (see Fig. 1.1), it briefly exceeded 2.7 m (8.9 ft) above *MLLW*. Note that the storm surge history (gray curve), though uneven over time, was very similar at both places attesting to the far ranging influence of the storm.

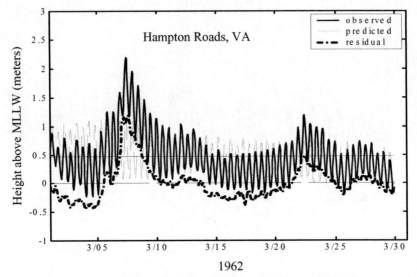

Fig. 9.3. Ash Wednesday storm tide recorded at Hampton Roads, VA.

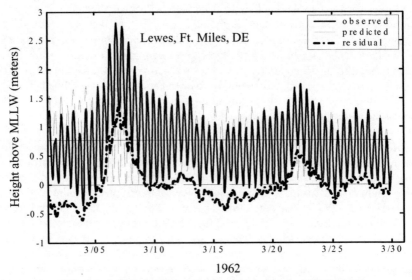

Fig. 9.4. Ash Wednesday storm tide recorded at Lewes (Ft. Miles), DE.

Hurricane *Isabel* – In mid-September, 2003, just when I thought I would never have another close encounter with a Category 5 hurricane, a storm named *Isabel* appeared in the western Atlantic bearing maximum surface winds of 140 knots (160 mph). Virginia had not experienced the full effect of a major hurricane since the year 1933 and it quickly became a matter of real concern when *Isabel*, slightly diminished in intensity but still producing winds of 135 knots (155 mph), began to track very decisively toward Chesapeake Bay. Few could be blamed for harboring secret hopes that its direction might change (it did not) but fewer still were prepared for what may have been the reprieve of the century – in the three days of her final approach toward land, *Isabel* traversed almost the entire Saffir-Simpson scale, from just below Category 5 all the way down to a Category 1 hurricane as it proceeded into Virginia and southern Chesapeake Bay after an initial landfall on the Outer Banks of North Carolina. Even so, high winds and high storm tides devastated the coastal region, particularly the tidal regions in the Bay and its tributaries. There was a reason for this.

Unlike previous hurricanes, a considerable amount of wind data was gathered during *Isabel's* final approach toward land, much of it by ground-based radar installations in addition to direct measurement by aircraft. A research team from NOAA's Atlantic Oceanographic and Meteorological Laboratory later compiled a detailed picture of her maximum surface winds at the time of landfall as shown in the upper right corner of Fig. 9.4. As this diagram illustrates, the highest winds occurred on the *right side* of the northwest moving eye. The schematic drawing on the left in Fig. 9.4 shows why – the counter-clockwise winds circulating around the storm center are directed onshore to the right of the eye and offshore to the left of it; since the entire storm system was moving landward, the speed of the storm's movement (about 8 knots) was added to the onshore wind speed (right side) and subtracted from the offshore wind speed (left side).

Advection from the stronger wind fields above the surface may modify this picture to some extent but this is the reason why higher winds, and higher storm surge, are almost always found on the right-hand side of a moving storm in the northern hemisphere. (Cyclonic winds turn the opposite direction, i.e., clockwise, in the southern hemisphere). In the case of *Isabel*, this meant her biggest punch would be reserved for the land and water areas on her right, including the Chesapeake Bay and its tributaries. Isabel caused 36 deaths and almost two billion dollars in damages to homes, businesses and public facilities in Virginia.

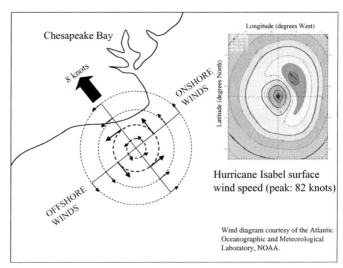

Fig. 9.4. Diagrams showing enhanced winds on the right side of Isabel.

Although *Isabel* destroyed a considerable amount of property in the Tidewater area of Chesapeake Bay including several tide gauges, the primary tide station for Hampton Roads survived and its record provides not only a glimpse of a large storm tide in the 21st century but allows a valuable comparison with one of the greatest storms of the 20th century in the US mid-Atlantic region: the hurricane of August 1933. A comparison of storm tide and storm surge for these two hurricanes is shown in Figs. 9.5 and 9.6.

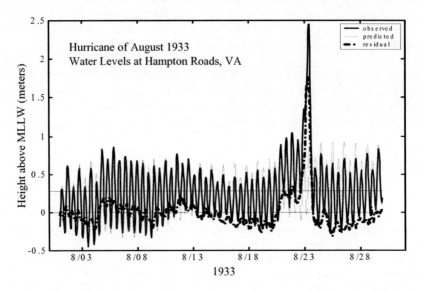

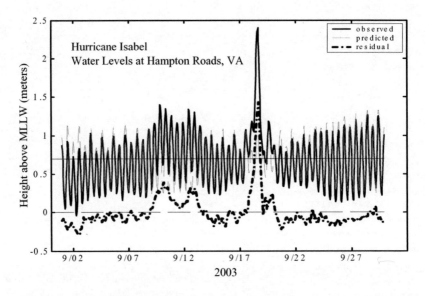

Fig. 9.5. Observed and predicted water levels at Hampton roads, VA, during August 1933, (upper panel) and September 2003 (lower panel).

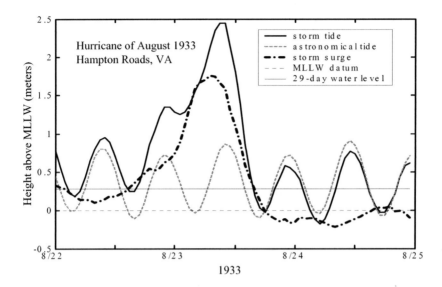

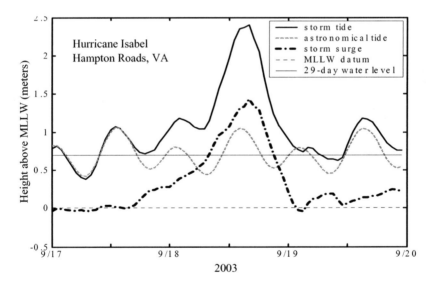

Fig. 9.6. Storm tide, astronomical tide, and storm surge at Hampton roads, VA, on August 23, 1933, (upper panel) and September 18, 2003 (lower panel).

Anatomy of a storm tide – The maximum storm tide recorded by the National Ocean Service during the August 1933 hurricane at Hampton Roads, Virginia, was 2.44 m (8.01 ft) above *MLLW*. The maximum storm tide recorded during hurricane *Isabel* at the same location was 2.40 m (7.87 ft) above *MLLW*. In light of the similarity between these numbers, it was not surprising to hear claims by some older residents that water marks left by *Isabel* were as high or higher than those seen on the same buildings after the 1933 storm. My own height estimate for the high water mark that I observed during *Isabel* came to about 2.5 m (8.1 ft) above *MLLW*, this along a sheltered bank behind my home on the York River.

The *storm surge* was another matter. The 1.78 m (5.8 ft) storm surge of the 1933 event at Hampton Roads clearly exceeded the 1.45 m (4.8 ft) storm surge produced by hurricane *Isabel* (Fig. 9.6). Why the difference? The August 1933 hurricane, now classified as a Category 3 storm, was more powerful in terms of its surge-generating potential. Hurricane *Isabel*, though a lesser storm, made up most of the difference because of a single factor: **sea level rise**. Astronomical tide range, so obviously important at Pensacola during *Camille* (Fig. 9.2), was not a major factor this time. *Isabel* took place during neap conditions while the 1933 storm occurred nearer to spring tides (Fig. 9.5), an advantage only slightly offset by the storm surge peaking nearer to the astronomical high tide during *Isabel* (Fig. 9.6). But as Fig. 9.5 shows, the **synodic (29-day average) water level** for September 2003 was significantly higher than the corresponding average during August 1933. The difference, about 0.41 m (1.3 ft), exceeds the 0.31 m (1.0 ft) that can be attributed solely to the secular sea level trend in Hampton Roads during the seventy years between 1933 and 2003[2]. But, as noted in Chapter 5 (Sec. 5.7, Fig. 5.3), monthly mean sea level variations above the secular trend in lower Chesapeake Bay can easily account for the 0.1m difference. We have an unlucky monthly mean sea level anomaly in addition to secular sea level rise to thank for *Isabel's* unusual storm tide.

A caution when observing storm surge - Although storm tide heights are measured directly by tide gauges, the 'observed' storm surge heights in Figs. 9.5 and 9.6 were obtained using the *superposition principle*; i.e., they were calculated as the difference between the storm tide and the astronomical tide derived from the 'best fit' of our astronomical tide model to 'time-local' water levels. Even assuming a perfect fit with all of the water level variance at tidal frequencies removed, errors may still arise at a critical point – a change in the phase (time of arrival) of the astronomical high tide nearest the peak surge for example – because of interaction between the two long waves involved: the storm surge and the astronomical tide. This can happen where non-linear, shallow-water effects are important as they often are for extratropical storms crossing the continental shelves of northwest Europe and the UK. These effects are usually not as important for tropical storms and hurricanes, especially those moving rapidly onshore on a track perpendicular to the coastline. The alternative is to use a mathematical model to generate storm surge hindcasts (see Sec. 9.6). We should keep in mind, however, that the storm surge output from these models is quite sensitive to wind stress magnitude and direction, driving parameters that must be carefully measured at a number of points spread over the model's operating domain. And before deciding on this approach, we might ask the modeler what was used to verify their model. In most cases, it's the same observed storm surge based on superposition.

---

[2] Based on the most recent NOAA/NOS data, sea level at Hampton Roads, VA is increasing by 4.25 mm/year (1.39 ft per century) relative to the land.

9.5 THE IMPORTANCE OF BEING REFERENCED

In the collection of figures presented above, the curves representing the predicted (astronomical) tide and the observed (storm) tide are **referenced water levels**. They happen to be referenced to *MLLW* (Mean Lower Low Water) in these examples for the convenience of plotting a time series with mostly positive numbers. They could be referenced just as easily to some other datum such as *MHHW* (Mean Higher High Water). Both datums are referenced in turn to an arbitrary but fixed level - the station datum or 'staff zero'. As explained in Chapter 5, the water level readings that US tide gauges actually record are heights above station datum. The station datum is fixed when the gauge is installed but tidal datums can only be determined after the station has been in operation for a time and a sufficient amount of data has been collected to determine them. After that, water level readings from the gauge can be corrected to reference *MLLW*, for example, by subtracting a single number from them: the height of the *MLLW* datum above the station datum.

The storm surge curve, on the other hand, is a **non-referenced water level**; it simply refers to a height measured above zero, the height of the still water level that exists (we don't know where exactly) in the absence of both the storm surge and the astronomical tide. Why is this important? Consider how many times you've heard on the evening news that the tide, due to the weather at hand, will be so many feet above normal. We really don't know where 'normal' is on the ground unless we have a referenced water level, such as the one referencing the predicted tide, to tell us. Otherwise, storm surge height is just another indicator of storm intensity – a quantitative measure like horsepower or dynamic pressure.

Referenced water levels are different. They can be measured and located on the ground as elevation contours forming a boundary, a valuable tool in establishing zones of storm tide risk. For that very important purpose, I believe the boundary of choice should be an extreme one such as the *mean higher high water* (*MHHW*) datum as explained in the following section.

The case for referencing storm tides to mean higher high water - Why measure a maximum storm tide height in units (feet or meters) above the tidal datum of *MHHW* rather than a low water datum? Basically we require the *MHHW* datum in order to isolate and evaluate storm surge risk in a conservative way by removing the effect of *tidal range* - an independent factor that varies from place to place. For example, US nautical charts use *MLLW* to reference charted depths conservatively so that a mariner will know that the water depths shown on the chart can be counted on for safe passage even at the lowest levels of the astronomical tide at the place in question. Reversing direction and looking upward instead of downward, *MHHW* can be used to conservatively reference storm tides so that coastal residents will know how much additional rise to expect above the highest levels of the astronomical tide. These levels are generally familiar to the waterfront resident who witnesses signs of their presence in wrack lines, marsh vegetation zones and high water marks on structures.

We'll sharpen this point with a pair of histograms combining *MHHW* elevations and storm tides of record for twelve NOS primary tide stations along the US Gulf and East Coasts (Fig. 9.7). 'Of record' means the highest water level recorded at the tide station during its period of operation. Of the twelve stations included, four had record storm tides due to Category 3 (*Alicia*) or Category 4 hurricanes (*Donna*, *Hugo*, and possibly an unnamed storm in 1926). But water levels referencing *MLLW* (black bars) show that the highest storm tide of record (7.4 m above *MLLW* at Eastport, ME) was due to a winter storm in January 1997. This storm tide appears overwhelming but is it? Because the New England coast (stations 11-12) has a large tidal range, the tidal datum of *MHHW* (gray bars) is also higher there as compared to the US Gulf coast (stations 1-3).

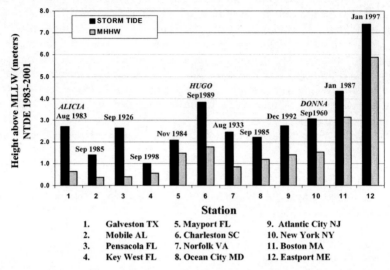

Fig. 9.7. Record storm tides measured above MLLW (1983-2001), US Gulf and Atlantic coasts.

The tidal range factor can be removed by referencing the same storm tides to *MHHW* as shown in Fig. 9.8. Now we see that five stations had equal or greater storm tides than Eastport, ME. All five experienced Category 4 hurricanes except Norfolk, VA (Category 3 in August 1933). It's very likely that Mobile Bay, AL also experienced a very large storm tide during the September 1926 hurricane but no tide station was in operation there at that time. The same may be true for other storms missed by the primary tide stations at Key West and Mayport, Florida.

To further appreciate the storm tide risk confronting the US Gulf coast states, we look again at sea level rise relative to the land and take note of the fact that storm tide heights recorded for older storms have been considerably 'shortened' in the interval since they occurred; i.e., the datums of reference have gone up dramatically along this rapidly sinking coast[3]. In making the regional comparisons shown in Figs. 9.7 and 9.8, I've employed a common datum based on the current US National Tidal Datum Epoch (1983-2001). This approach, while necessary to compare storm tides of record on the same scale, reflects the sea level conditions that existed at the time; if some of the older storms were to occur again today, they would clearly produce higher storm tides relative to the tidal datums currently in use.

---

[3] Recent NOAA/NOS data indicate that sea level at Galveston, TX is increasing by 6.50 mm/year (2.13 ft per century) relative to the land.

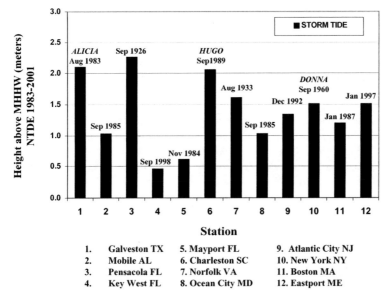

Fig. 9.8. Record storm tides measured above *MHHW* (1983-2001), US Gulf and Atlantic coasts.

By the way, there's another reason for choosing a more extreme tidal datum, *MHHW* or *MLLW* as opposed to *MHW* or *MLW*. In the 19-year interval defining *MHW* and *MLW*, two high waters and two low waters per tidal day are included in the averaging *unless* there is only one high and one low as happens during equatorial tides in regions where the tidal type is mainly diurnal (see Chapter 5). For *MHHW* or *MLLW*, only the one tide – the higher high or lower low - is used per tidal day throughout. For this reason, *MHW* and *MLW* tidal datums often are not continuous – they undergo a sudden jump in elevation when passing from a diurnal to a semidiurnal tide region because of the 'missing' tides in the diurnal region.

Reasons for differences in tide range along the US East Coast – The gray bars were introduced in Fig. 9.7 to give an idea of the wide variation in tidal range for the waters between the US Gulf of Mexico and the Gulf of Maine. Because they display the difference between mean higher high water and mean sea level, the height of these bars approximates one half of the *diurnal tide range (MHHW-MLLW)*. The diurnal range remains small in the Gulf of Mexico due to its separation from its parent tide regime in the Atlantic (Chapter 3, p.43-44). Variations in diurnal range along the Atlantic coast, however, are largely a function of the varying width of the continental shelf (Chapter 6, p.73). The shelf zone can be considered to be an elongated, co-oscillating tidal basin closed at its shoreward end. Comparing east coast shelf widths (as surrogates for basin length) that vary between roughly 50 and 250 km to tidal wavelengths on the order of 1000 km for an average 50 m shelf depth, we are mostly inside the one-quarter 'resonance' wavelength (see Fig. 6.8) where tidal range at the coast should increase with shelf width, other factors being equal. Wide shelves also facilitate greater storm surge height but in place of a regular tide wave, we have a long wave that is generated by a single moving disturbance: the tropical storm or hurricane.

9.6 PREDICTING STORM SURGE AND STORM TIDE

It is a complex picture that unfolds as we follow the news reports and view the satellite images of a major hurricane. The life of a typical storm event last several days from the first signs of a tropical depression developing into a tropical storm and finally a hurricane as the storm grows in strength and steadily tracks its way across the tropical ocean towards land. During that time the storm system may change direction, increase, or slow its forward progress in an almost whimsical manner. But as far as the storm surge is concerned, most of the action occurs within 48 hours or less as the storm crosses a relatively shallow zone - the continental shelf – lying between the shoreline and the shelf break, a transition into deep water outlined by the 100-meter depth contour. As the hurricane progresses inside the shelf break, the surge progressively builds and may continue to grow to dangerous heights, particularly if the shelf is a wide one and the storm is moving at optimum speed. Comparing and contrasting the respective forces of a hurricane and the astronomical tide can provide some insight into this process.

Unlike the tide-producing forces (weak but continually acting on all water masses in the world's oceans), the forces imparted by a hurricane are immense but locally concentrated on the sea surface beneath the central region of the storm. Through wind stress acting directly on the water's surface, high momentum from the moving air in the atmosphere is transferred to the water column, setting water in motion across the shelf that leads to a setup or general rise in water level. Changes in atmospheric pressure and the generation of a long wave propagating in the direction of the storm can add to the setup. Let's look at these last two factors in more detail.

All storms are characterized by low atmospheric pressure. Hurricanes in particular show a large drop in pressure near the eye. At its landfall near Bay St. Louis, Mississippi, hurricane *Camille* had a recorded atmospheric pressure of 681.7 mm (26.84 inches) of mercury compared to a normal atmospheric pressure of 780.0 mm (29.92 inches) of mercury at sea level. This is the pressure measured with a mercury barometer - basically a long tube sealed at one end, filled with mercury, and then inverted in a dish of mercury. Decreasing atmospheric pressure on the mercury in the dish causes the mercury in the tube to fall. If in place of a column of mercury (a metal of very high density) we substituted a column of seawater in an enlarged barometer, this would correspond to a drop from 10.1 m to about 9.1 m of normal seawater as read on the scale of the barometer. However, the ocean's surface responds opposite to the way a barometer does and this is called the *inverted barometer effect*. Hence the atmospheric pressure drop in this case would cause a water level increase of about 1 meter. Obviously this is just one part of the total setup experienced during *Camille*.

The moving disturbance created by low atmospheric pressure and the belt of maximum winds surrounding the eye of a hurricane creates a *long wave* moving across the shelf. The amplitude of this wave can become large *if* the hurricane has the right forward speed for the water depth. Why is this? As explained in Chapter 3, long waves are shallow water waves that move at a speed (celerity) governed by the water depth as given by the formula $C = \sqrt{gh}$. For example, suppose the water depth, h, at some point happens to be 20 m. According to this formula, a long wave there would have a celerity, C, of 14 m/sec or about 31 miles per hour. If a hurricane should happen to be moving landward over the same depth at that speed (a very fast hurricane), the wave and the storm become synchronized. This allows the storm to transfer more of its energy to the wave, continually building the wave's amplitude as the two move shoreward together. However, should the storm be moving faster or slower than the *critical speed* for that depth, long waves of lesser amplitude continually form and propagate away from the storm center as free waves.

Mathematical models – Until the late 1960's, researchers relied on perceived cause-and-effect relationships and sparse data sets to develop empirical formulas for predicting setup, adding them together to estimate storm surge. With the possible exception of the atmospheric pressure drop and the inverted barometer effect, each relationship had its restrictions and no formula could account for all of the complexities of real hurricanes and tropical storms.

As a result of advances in computer technology, the empirical approach has since been replaced by mathematical model predictions that offer important advantages. Unlike empirical formulas that were assumed to have universal application, each mathematical model provides customized results for a given shelf area and section of the coastline – the *model domain*. Rather than make predictions at a single point, these models deliver predictions at thousands of points within an array of cells comprising the model grid.

Fig. 9.9 represents a computer-drawn image showing the highly localized storm surge generated in Biscayne Bay during hurricane Andrew in 1992. Post-storm simulations were made with one of the newer mathematical models known as U-TRIM. As the figure illustrates, storm surge varies greatly from point to point inside an embayment or within regions of the coast with complex nearshore bottom topography guiding the approaching storm wave.

Fig. 9.9. Storm surge simulation for Biscayne Bay, Southeast Florida, during hurricane Andrew using the UnTrim Mathematical Model (courtesy of Dr. Jian Shen, Virginia Institute of Marine Science). Areas of highest surge appear just below south Miami (maximum height: 5 m).

Clearly some very valuable information is available to coastal emergency planners assuming model forecasts can be completed far enough in advance to provide adequate warning for the specific areas requiring evacuation during storms. The quality of model predictions continues to improve in a number of ways. The two most notable are: (1) improvements regarding the math behind the model – the computer algorithms and the hydrodynamics they support; (2) improvements in the data used to construct, calibrate, and verify the model. For the second line of improvement one can now add thousands of water depths digitized and located spatially with extreme precision using GIS (Graphical Information Systems).

9.7 STORM TIDES AND SEICHES

A periodic surge called a **seiche** occurs in semi-enclosed seas and gulfs that, in close accord with their dimensions, respond to meteorological forcing in one or more resonant modes (see Chapter 3, Sec. 3.6). In the Adriatic Sea, seiches can be very pronounced and rival the astronomical tide when triggered by strong weather events. In addition to seiches with well-defined periods of oscillation lasting several days, the shallow northern section of the Adriatic is also subject to transient storm surge events; the combination of transient storm surge, seiche and astronomical tide poses a special threat to the fabled city of Venice that lies exposed to all three inside a shallow lagoon at the head of the Adriatic[4].

Two dominant wind systems, the *bora* and the *sirocco*, affect the Adriatic Sea. The bora represents a strong outpouring of polar air entering the Adriatic through gaps in the Dinaric Alps on its eastern shore.

Fig.9.10. Bathymetric relief map of the Adriatic Sea showing the location of Venice, Italy, at its northern boundary.

---

[4] A series of underwater floodgates have been proposed at the entrances to the lagoon to protect Venice from frequent flooding due to storm tides arising in the Adriatic Sea. In 1997, the city experienced about 100 separate instances of flooding.

Sirocco winds are tropical in origin. They begin with warm air from North Africa and Arabia moving in frontal systems to the northwest, producing high velocity southeasterly winds parallel to the main axis of the Adriatic Sea. Together with bora winds, they create a formidable mechanism for generating both storm surge and seiche. Typically, as a cyclonic low approaches the Adriatic, sirocco winds add surface stress to the effect of low atmospheric pressure, causing water to pile up at its northern end. With the passage of the low, **sirocco winds** are succeeded by **bora winds** from the transverse direction that further assist in developing and prolonging the seiche.

Perhaps the most striking example of this surge-seiche scenario is the record storm of November 4, 1966. Fig. 9.11 shows the resulting water level history over a 15-day period at the Punta Salute tide station in Venice, Italy. The reference datum in this case is a mean sea level determination known as the IGM, an average based on the years 1937-1946. Further discussion of this datum and the sea level trend at Venice follows momentarily.

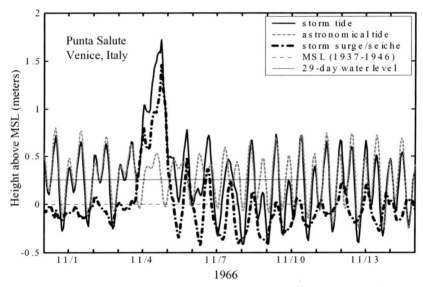

Fig. 9.11. Record storm tide measured at Venice, Italy, on November 4, 1966. The peak level recorded at Punta Salute was 1.72 m above MSL (IGM). Data provided courtesy of the Venice Tide Forecasting Centre.

The impressive storm surge revealed in Fig. 9.11 begs for comparison. Comparing the 1966 storm at Venice with hurricane Isabel in Virginia (Fig. 9.6), we see that the storm surges in both instances reached a maximum of about 1.5 m[5]. In that respect, we may say that the 1966 storm at Venice was the equivalent of a hurricane! But here the

---

[5] Surge heights are determined here as the difference between the storm tide and the astronomical tide, two locally referenced water levels (Sec. 9.6). Surges are universal: only the zero (dotted) line is needed to compare the surge heights in Figs. 9.6 and 9.11.

comparison ends. The seiche that followed the maximum storm tide at Venice has no counterpart in Chesapeake Bay.

The clue to understanding what is unique about the seiche in Fig. 9.11 can be found in the periodogram for the residual levels at Venice in November 1966. In addition to a storm-induced cascading of energy toward zero frequency, Fig. 9.12 shows a sharp peak at a mean frequency of 1.118 cpd and a lesser peak at a mean frequency of 2.224 cpd. These frequencies correspond to periods of 21.5 hours and 10.8 hours respectively, representing the first and second modes of the well-known northern Adriatic seiche, a phenomenon carefully studied for a number of years. Italian scientists in a 1983 study listed periods of 21.4, 10.8, 7.2 and 5.3 hours for the four seiche modes observed on the Adriatic shelf fronting the Venetian lagoon.

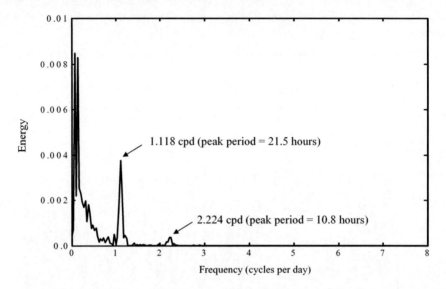

Fig. 9.12. Periodogram of the tidal residual at Punta Salute, Venice, Italy, during November 1966.

Curiously, the analyses reported for the northern Adriatic suggest a gradual decrease in period of the fundamental seiche mode from 23 hours to 21 hours between 1910 and 1980, a perceived change believed to be due to variations in analytical techniques rather than changing bathymetry throwing the instrument (Adriatic Sea) out of tune. Although the fundamental mode is near that of a minor diurnal tide constituent (OO_1 - period 22.3 hours), none of the higher modes appear to coincide with tidal frequencies.

Clearly one of the most impressive attributes of a seiche is the tide-like precision of its modal frequencies – frequencies matching the water body and not the storm (e.g., the 4.8 day sub-tidal oscillations induced by extratropical storms in Chesapeake Bay, Fig. 7.3). However, while its frequency may be quite regular, the phase of a group of seiche oscillations raises several questions: Is the initial phase determined by the timing of the triggering event? Does it remain steady? Is there interaction with the astronomical tide? To explore the last of these questions we should take a closer look at the first few seiche oscillations following the November 4 storm surge at Venice as they appear on the right in Fig. 9.13 below.

Even without recourse to more sophisticated analytical techniques including cross-spectral analysis, we notice something peculiar about the seiche in Fig. 9.13: it has lows that consistently coincide with astronomical highs while its highs tend to fall at mid-tide or lower levels (note that the type of tide at Venice is mixed, mainly semidiurnal). Other examples not shown yield similar results, suggesting that there is an interaction mitigating against the highest water levels that could potentially emerge from a combination of seiche and astronomical tide fully in phase. If true, it's good news for the city of Venice since the seiche amplitude can be as large or larger than that of the astronomical tide, a fact quite apparent in Fig. 9.13.

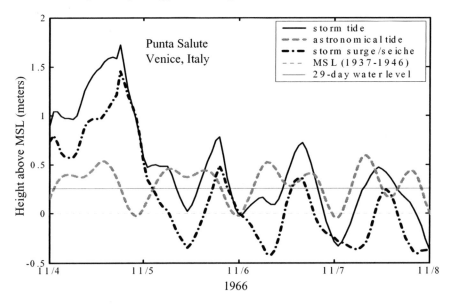

Fig. 9.13. Record storm tide and seiche measured at Venice, Italy, on November 4, 1966. Data provided by the Venice Tide Forecasting Centre.

Is Venice sinking? – We cannot discuss Venice and its flooding problems without at least a brief mention of the sea level trend observed there. As always when the focus is on flood risk locally; the only sea level trend we really care about is the one relative to the land. According to the National Research Centre of Italy, pumping of groundwater for industrial purposes caused the land around Venice to subside by approximately 23 cm between 1920 and 1970, an apparent water rise of 4.6 mm/yr. Since this practice was stopped in 1970, regional subsidence appears to have ceased. But local subsidence is only one issue. What about the larger picture? For that we must balance the *eustatic* or worldwide rise in sea level against ongoing vertical movements by the earth's lithospheric 'plates' (see Chapter 5, Sec. 5.7). There appear to be happy circumstances associated with the plate Venice is riding on – or so the water level data suggest.

Fig. 9.14 contains a plot of yearly mean sea level (average of hourly heights) for thirty-one years at the Punta Salute tide station following 1970 when pumping was stopped. No trend is apparent in the figure and the Venice authorities, who continue to use the 1937-1946 IGM datum as their sea level reference, have reported none. It's hard

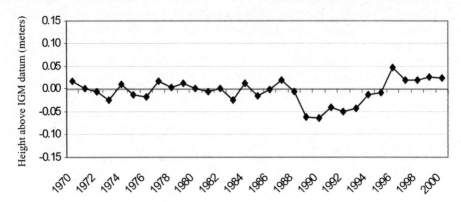

Fig. 9.14. Yearly mean sea level, 1970 through 2000 at Punta Salute tide station, Venice, Italy; Data provided by the Venice Tide Forecasting Centre.

to argue against this view based on the years 1970-1988 but one wonders whether an upward trend may have re-emerged starting in 1989. A related factor with an impact on flooding is the seasonal variation in water level. Fig. 9.15 contains a plot of monthly mean sea level adjusted for (computed as a deviation from) the annual means for the years 1970 through 2000 shown in Fig. 9.14. The solid curve in Fig. 9.15 gives the expected systematic variation in water level about each yearly mean (the data needed to determine the seasonal tide constituents Sa and Ssa) while the confidence intervals (vertical lines) show the expected random variation about each monthly mean. The data suggest that the seasonal tide cycle offsets surge/seiche flooding in January, February and March while adding to it in October, November and December. Plus the random anomalies appearing in these months are likely to be far greater than at other times of the year. November 1966 appears to have been one of those unfortunate months.

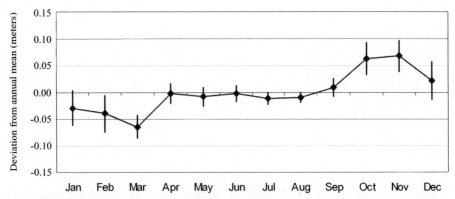

Fig. 9.14. Monthly mean sea level, 1970 through 2000 at Punta Salute tide station, Venice, Italy. Vertical lines through means indicate 95% confidence interval. Data provided by the Venice Tide Forecasting Centre.

10

Computational methods: the matrix revisited

10.1 ADVANTAGES OF MATLAB® MATRIX ALGEBRA

Matrix algebra provides a concise way of handling an array of numbers and extracting useful information from it. Like simple arithmetic, working with matrices can be daunting if we're forced to do it 'by hand'. But, just as few of us would bother to work out the square root of ten with pencil and paper (we'll simply enter ten on our handheld calculator and hit the square root key), it's possible to do quite a lot with matrices just by entering simple commands in MATLAB[1], a computer programming language that relies on matrix algebra to perform a broad range of math operations. After a brief review of some basic matrix operations, we'll see how they are applied to the analysis of tides and currents using Harmonic Analysis, Method of Least Squares (HAMELS).

The matrix - When data are arranged in an array with rows and columns of numbers, it's called a matrix. One of the first things we need to know is the *size* of the matrix – how many rows and how many columns it contains. If it contains only one row *or* one column, it's called a **vector**. If it contains only one row *and* one column, it's a **scalar**. A typical array of water level data might form a 696 by 2 matrix - we'll call it X - that contains 696 rows of time and height values (696 consecutive hours in column one and the corresponding hourly heights in column two). If we've loaded a water level time and height series into X, we can confirm its size by entering the following expression on MATLAB's command line:

$$[n\ m] = size(X)$$

After pressing the ENTER key, the response will be $n = 696$, $m = 2$. In general, a matrix with n rows and m columns is referred to as an $n \times m$ matrix with $n \times m$ numbers *or elements*. To indicate a particular element in the matrix, we simply specify the numerical indices identifying the row and column; for example, the third recorded time in X would be the element $X(3,1)$ and the corresponding height would be the element $X(3,2)$. In a typical MATLAB program, $X(i,j)$ identifies the element in the ith row and jth column.

Matrix addition and subtraction - The reason why matrix size is important becomes clear when we perform a math operation involving two or more matrices. To perform addition or subtraction of two matrices, X and Y, the matrices must be identical in size so that their corresponding elements can be added or subtracted. To perform addition in MATLAB, we would write

---

[1] The Math Works, Inc., 3 Apple Hill Drive, Natick, MA, 07160-2098 USA

$$Z = X + Y$$

and obtain the sum in a new matrix, Z (An error occurs if X and Y are *not* the same size). We could also write

$$Z = Y + X$$

to obtain the same answer, meaning that the order of matrix addition and subtraction does not matter. We simply add (subtract) the corresponding elements in each matrix.

Matrix multiplication – To multiply two matrices together and obtain their product, the matrices must be **conformable**. This means that the number of columns in the first matrix entered must equal the number of rows in the second; put another way, there must be a *square inner product* (2 x 2, 4 x 4, etc). For example, suppose that X is a 3 x 2 matrix and Y is a 2 x 4 matrix. In MATLAB, we could then write

$$Z = X * Y$$

On paper, however, I might first check to see if the matrices are conformable by writing

$$\begin{matrix} 3 \times 4 & 3 \times 2 & 2 \times 4 \\ [Z] & = [X] & [Y] \end{matrix}$$

In the above equation, Y is *pre-multiplied* by X and this is valid because the inner product is square (2 x 2). The outer or final product is 3 x 4, the size of the new matrix Z. However, X pre-multiplied by Y would be invalid since the inner product would then be non-square (4 x 3). Clearly the order of matrix multiplication is important.

You may have noticed that I enclosed the matrix symbols in brackets in the last equation. MATLAB allows you to use either form of representation but the brackets are useful when you want to show each element in a matrix equation like the following one,

$$\begin{bmatrix} 55 & 80 \\ 80 & 230 \end{bmatrix} = \begin{bmatrix} 1 & 2 & 3 & 4 & 5 \\ 6 & 7 & 8 & 9 & 0 \end{bmatrix} \begin{bmatrix} 1 & 6 \\ 2 & 7 \\ 3 & 8 \\ 4 & 9 \\ 5 & 0 \end{bmatrix}$$

Here we have a 5 x 2 matrix pre-multiplied by a 2 x 5 matrix yielding a 2 x 2 matrix as a result. We get the elements shown in the 2 x 2 product matrix by *row-column inner product summation*. You and I wouldn't want to get stuck with it but your computer loves to do math this way. Multiplying in turn the five elements in the first *row* of the left matrix by the five elements in the first *column* of the right matrix and summing the products yields 55, the number in the first row, first column of the product matrix; i.e.,

$$1^2 + 2^2 + 3^2 + 4^2 + 5^2 = 55$$

Performing the same operation with the first row of the first matrix and the second column of the second matrix yields 80, the number in the first row, second column of the product matrix; i.e.,

$$1 \cdot 6 + 2 \cdot 7 + 3 \cdot 8 + 4 \cdot 9 + 5 \cdot 0 = 80$$

The product matrix in this case is a **square matrix** with an equal number of rows and columns – size $n \times n$ in other words. A square matrix has *diagonal elements* – elements with equal row-column indices - running downward and across the matrix from left to right. The diagonal elements in the square matrix above are 55 and 230. The *off-diagonal elements* (those appearing above and below the diagonal) are both 80, meaning that this particular square matrix is also a **symmetric matrix**.

The example just given has considerable practical value. The 5 x 2 matrix on the right has the form of a *bivariate data set* – two variables measured on five objects or samples; e.g., five consecutive samples of time and water level height. The 2 x 5 matrix on the left is the *transpose* of the 5 x 2 matrix on the right, meaning it's the same matrix but with the rows and columns switched. We will always get a square symmetric matrix when we pre-multiply a given matrix by its transpose. In MATLAB, the transpose is indicated by an apostrophe entered after the matrix symbol. For example:

$$Z = X'*X$$

Pre-multiplying a bivariate data matrix by its transpose conveniently yields the sums of squares and sums of cross products terms that represent one of the steps in calculating *variance* and *co-variance* for two or more variables. These fundamental quantities are required in certain statistical tests and in other numerical procedures routinely performed with **multivariate data**.

Matrix division – In ordinary algebra, we can divide a number by any other number except zero. In matrix algebra, division *per se* is not defined although *matrix inversion* is. So instead of dividing 6 by 2, for example, we have to perform the equivalent operation and multiply 6 by ½, the inverse of 2. However inversion applies only to square matrices and not all of these have an inverse (a matrix that has no inverse is called a **singular** matrix). Just as the inverse of 2 can be written as 2^{-1}, the inverse of a matrix X is often indicated by the symbol X^{-1}.

The steps involved in matrix inversion are beyond the scope of this book but again it's a simple job for your computer. Using MATLAB, you can get the inverse of a square matrix X in two ways: with exponential symbols, $X^{\wedge -1}$, or by using a command word and argument:

$$inv(X)$$

Either one will produce the inverse of X, or else give you a warning if the matrix is singular or nearly so. You can check the result by pre-multiplying X by its inverse, $X^{-1}X$. This is like dividing a number by itself – you should get unary one. Only here you will get the **identity matrix,** I, where

$$I = inv(X)*X$$

An identity matrix is the matrix equivalent of unary one. All of its diagonal elements are ones and all of its off-diagonal elements are zeroes. Multiplying I by any other conformable matrix has no effect on that matrix.

Simultaneous equations in matrix form – Simultaneous equations lie at the heart of many numerical procedures including regression analysis and least squares harmonic analysis. In a multiple linear regression problem, we might have three *independent* variables (X_1, X_2, X_3) that we want to use in linear combination to predict a *dependent* variable Y. Making the minimum set of three observations then leads to a set of three simultaneous equations in three unknowns

$$a_1 X_{11} + a_2 X_{12} + a_3 X_{13} = Y_1$$
$$a_1 X_{21} + a_2 X_{22} + a_3 X_{23} = Y_2$$
$$a_1 X_{31} + a_2 X_{32} + a_3 X_{33} = Y_3$$

These equations can be written in matrix form as:

$$\begin{bmatrix} X_{11} & X_{12} & X_{13} \\ X_{21} & X_{22} & X_{23} \\ X_{31} & X_{32} & X_{33} \end{bmatrix} \begin{bmatrix} a_1 \\ a_2 \\ a_3 \end{bmatrix} = \begin{bmatrix} Y_1 \\ Y_2 \\ Y_3 \end{bmatrix}$$

The solution for the unknowns is found by pre-multiplying both sides of the above equation by the inverse of the square (independent variable) matrix. This will yield a vector of unknowns

$$\begin{bmatrix} a_1 \\ a_2 \\ a_3 \end{bmatrix} = \begin{bmatrix} X_{11} & X_{12} & X_{13} \\ X_{21} & X_{22} & X_{23} \\ X_{31} & X_{32} & X_{33} \end{bmatrix}^{-1} \begin{bmatrix} Y_1 \\ Y_2 \\ Y_3 \end{bmatrix}$$

Armed with these basic tools from matrix algebra, we can handle with ease certain mathematical operations that posed a far greater challenge only a few decades ago to anyone engaged in the analysis of time series data – data that include water levels and water currents. However, two approaches exist that require some understanding before we turn to the subject of tide and current analysis by the method of least squares.

10.2 HARMONIC ANALYSIS COMPARED WITH SPECTRAL ANALYSIS

To practitioners of conventional time series analysis, it might seem that spectral analysis of the type commonly found in engineering software packages under the heading of signal processing would be ideal for tidal analysis. This is not the case because of the way spectral frequencies are handled. A brief description of spectral analysis, its strengths and limitations is in order before introducing harmonic analysis and the least squares procedures employed in this book.

Spectral analysis - The aim of most spectral methods is to analyze the energy spectrum of a signal after transformation from the **time domain** to the **frequency domain**. After sampling the sinusoidal cycles making up the signal as they oscillate through time (time domain), we use spectral analysis to find the distribution of signal variance - the *energy* associated with each cycle at an assigned frequency (frequency domain). Total energy is expressed as the total variance or *mean square value* of the signal - usually a discrete time series in the form of a vector x_t of length n and time step (sampling interval) Δt.

Variance about the series mean value, $\bar{x}$, is calculated as $\sum_{i=1}^{n} (x_i - \bar{x})^2 / n$, a sum that can be partitioned through spectral analysis into n/2 contributing mean square components associated with n/2 discrete **Fourier frequencies**, ω_j, where

$$\omega_j = 2\pi j / n\Delta t \; ; \; j = 1...n/2$$

is the jth Fourier frequency expressed in radians per unit time. Note that all of the Fourier frequencies are multiples of the fundamental frequency ($j = 1$). The 'zeroth' Fourier frequency ($j = 0$) is associated with the series mean.

Series duration and frequency bandwidth – Suppose we happen to have a standard time series comprised of 696 hourly water levels that we want to analyze. Since the time increment and length of the series are already determined, there are 348 ordered frequencies with spectral energy estimates, each one separated from its neighbors by a predetermined *frequency bandwidth* value $\Delta f = 1/n\Delta t = 1/T$ where T is the series duration or **fundamental period**. Obviously, by increasing the duration of the series, bandwidth may be reduced. This is desirable for two reasons: one, it allows better resolution of energy peaks that happen to fall close to one another in frequency; two, the relatively large error associated with energy estimates at individual Fourier frequencies can be improved by 'smoothing' - averaging the estimates over one or more adjacent frequencies from a single analysis (bandwidth averaging) or averaging the corresponding estimates from several separate analyses (ensemble averaging).

Time step and aliasing - A vexing problem called **aliasing** can occur if the time step, Δt, is allowed to exceed a certain limit. The limit appears in the **Nyquist frequency** or 'cutoff' frequency defined as $f_c = 1/(2\Delta t)$. Once a particular time step is used to obtain discrete samples of a continuous time signal, say $\Delta t = 0.5$ hours, we risk data aliasing if that signal contains energy at frequencies above $f_c = 1/(2\Delta t) = 1$ cycle per hour. Fig. 10.1 shows that 0.5-hour sampling of a signal oscillating at that frequency might sample the highs and lows correctly, yielding a saw-tooth wave with the correct amplitude, or it might sample the midlevel of the curve yielding a flat line. But if the sampling rate is increased beyond $\Delta t = 0.5$ to, say 0.83 hours, a wave with a much lower frequency – 0.2 cycles per hour – is represented in our sample.

When spectral analysis is performed, the energy from the false (aliased) signal duly appears in the frequency band centered on 0.2 cycles per hour and not the correct 1 cycle per hour. Once a continuous signal has been sampled in this way, the resulting data are permanently compromised. Either a shorter Δt or special smoothing techniques, e.g., low-pass filtering[2], must be applied beforehand to avoid this difficulty. Both harmonic and spectral analysis require positive steps to avoid the aliasing problem.

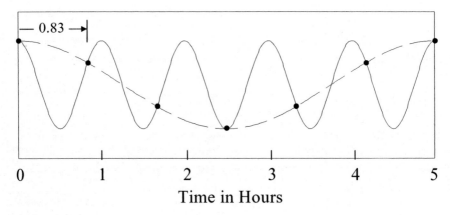

Fig. 10.1. Discrete sample aliasing of a cosine wave, $f = 1$ cycle per hour.

Spectral leakage - As noted above, a longer series improves resolving power or the ability to discriminate between adjacent energy peaks in the frequency spectrum. However, the key problem, as far as tidal analysis is concerned, is that a wave oscillation within the signal whose frequency happens to fall between two Fourier frequencies distributes its mean square contribution not only to these two but also to other adjacent frequencies through the phenomenon of *leakage*. Even with proper signal conditioning and smoothing (bandwidth averaging or ensemble averaging) leakage is never entirely eliminated and the error associated with spectral energy estimates at individual frequencies remains relatively high. This is especially true for short series of

---

[2] A tide gauge stilling well serves as a mechanical filter of this type.

so-called geophysical data. While spectral analysis is the proper choice when the goal is to investigate the unknown distribution of energy over a full range of frequencies, there's an easier way when the energy we seek occurs at just a few frequencies known in advance.

Harmonic analysis - Harmonic analysis takes advantage of the fact that, unlike many other time signals that exist in the natural world, tides and tidal currents are forced oscillations that occur, as explained in Chapter 4, only at known tidal frequencies. Depending on the hydrodynamics of the region, more than a hundred *tidal constituents* (sinusoidal waves with a constant frequency, amplitude, and phase) may exist. But, as in regression analysis, we view these as a collection of independent variables and care only about the ones that make dependent variable predictions work. In many areas, harmonic tide models frequently account for 80 to 90 percent of the variance in recorded water levels using only eight or nine of the major tidal constituents. And with harmonic analysis, you get to choose the frequencies. With spectral analysis, you get the Fourier frequencies.

10.3 HARMONIC ANALYSIS, METHOD OF LEAST SQUARES (HAMELS)

The least squares method used in tidal analysis can be viewed as a form of multiple linear regression. To see how it applies to tidal analysis, we will first consider another example of multiple regression: the so-called 'predictor' equation in three unknowns with two independent variables

$$Y = a_0 + a_1 X_1 + a_2 X_2 + E \quad (10.1)$$

In Eq. 10.1, two independent variables, X_1 and X_2, are linearly combined to provide an estimate, $\hat{Y}$, of a dependent variable, Y, their difference being the estimation error, $E = Y - \hat{Y}$. The central task of the least squares procedure is to find the set of unknowns ($a_0, a_1, a_2..$) that minimizes the squared error function

$$\sum_{i=1}^{n}(Y_i - \hat{Y}_i)^2 = \sum_{i=1}^{n} E_i^2$$

given a data vector, Y_i, $i = 1..n$. The function is minimized by taking partial derivatives with respect to the unknowns and setting them equal to zero. Omitting the summation indices, this takes the form

$$\frac{\partial \sum E^2}{\partial a_0} = \frac{\partial \sum E^2}{\partial a_1} = \frac{\partial \sum E^2}{\partial a_2} = 0$$

$$\sum E \frac{\partial E}{\partial a_0} = \sum E \frac{\partial E}{\partial a_1} = \sum E \frac{\partial E}{\partial a_2} = 0$$

Introducing the error term

$$E = Y - \hat{Y} = Y - (a_0 - a_1 X_1 - a_2 X_2)$$

then yields a set of normal equations

$$-\sum (Y - a_0 - a_1 X_1 - a_2 X_2)\ 1 = 0$$
$$-\sum (Y - a_0 - a_1 X_1 - a_2 X_2) X_1 = 0$$
$$-\sum (Y - a_0 - a_1 X_1 - a_2 X_2) X_2 = 0$$

Converting the equations to matrix form

$$\begin{bmatrix} n & \sum X_1 & \sum X_2 \\ \sum X_1 & \sum X_1^2 & \sum X_1 X_2 \\ \sum X_2 & \sum X_1 X_2 & \sum X_2^2 \end{bmatrix} \begin{bmatrix} a_0 \\ a_1 \\ a_2 \end{bmatrix} = \begin{bmatrix} \sum Y \\ \sum X_1 Y \\ \sum X_2 Y \end{bmatrix}$$

and pre-multiplying both sides by the inverse of the 3 x 3 matrix above,

$$\begin{bmatrix} a_0 \\ a_1 \\ a_2 \end{bmatrix} = \begin{bmatrix} n & \sum X_1 & \sum X_2 \\ \sum X_1 & \sum X_1^2 & \sum X_1 X_2 \\ \sum X_2 & \sum X_1 X_2 & \sum X_2^2 \end{bmatrix}^{-1} \begin{bmatrix} \sum Y \\ \sum X_1 Y \\ \sum X_2 Y \end{bmatrix}$$

or, as expressed using matrix symbols,

$$[A] = [SSX]^{-1}[SXY] \qquad (10.2)$$

In place of a plain vanilla multiple linear regression like that of Eq. (10.1), tidal analysis for m tidal constituents requires a slightly more sophisticated linear equation representing periodic motion with sine and cosine terms; i.e.,

$$h(t) = A_0 + \sum_{j=1}^{m} A_j \cos \omega_j t + \sum_{j=1}^{m} B_j \sin \omega_j t \qquad (10.3)$$

Here the unknowns are the coefficients $A_0, A_1, B_1, ..., A_m, B_m$ and t is serial time. Note there is a mean but no aperiodic trend in the equation. Because Eq. (10.3) is a linear equation, its least squares solution has the same form as Eq. (10.2). To demonstrate this considering only the first three terms in Eq. (10.3), we first define

$$[X] = \begin{bmatrix} 1 & \cos\omega_1 t_1 & \sin\omega_1 t_1 \\ 1 & \cos\omega_1 t_2 & \sin\omega_1 t_2 \\ 1 & \cos\omega_1 t_3 & \sin\omega_1 t_3 \\ .. & .. & .. \\ 1 & \cos\omega_1 t_n & \sin\omega_1 t_n \end{bmatrix} ; \quad [Y] = \begin{bmatrix} h_1 \\ h_2 \\ h_3 \\ .. \\ h_n \end{bmatrix}$$

and then pre-multiply [X] by its transpose to obtain the first matrix on the right in Eq. (10.2):

$$[SSX] = [X]'[X] = \begin{bmatrix} n & \sum_{i=1}^{n} \cos\omega_1 t_i & \sum_{i=1}^{n} \sin\omega_1 t_i \\ \sum_{i=1}^{n} \cos\omega_1 t_i & \sum_{i=1}^{n} \cos^2\omega_1 t_i & \sum_{i=1}^{n} \sin\omega_1 t_i \cos\omega_1 t_i \\ \sum_{i=1}^{n} \sin\omega_1 t_i & \sum_{i=1}^{n} \sin\omega_1 t_i \cos\omega_1 t_i & \sum_{i=1}^{n} \sin^2\omega_1 t_i \end{bmatrix}$$

obtaining the second one after pre-multiplying the matrix [Y] by [X] transpose,

$$[SXY] = [X]'[Y] = \begin{bmatrix} \sum_{i=1}^{n} h_i \\ \sum_{i=1}^{n} h_i \cos\omega_1 t_i \\ \sum_{i=1}^{n} h_i \sin\omega_1 t_i \end{bmatrix}$$

The presence of so many sine and cosine product and summation terms in matrices [SSX] and [SXY] is the reason spectral analysis tended to slow computers of an earlier age to a crawl. Although the square matrix [SSX] can be readily calculated with the aid of certain orthogonal properties of the transcendental functions, inverted, and stored in advance, the vector matrix [SXY] must be calculated for each new data set at the time of analysis. Inserting a set of $n/2$ Fourier frequencies in place of the single frequency, ω_1, in the example above, the calculation of [SXY] would come at a cost of n^2 multiplications and additions in the *cosine and sine **Fourier transform** terms*, $\sum_{i=1}^{n} h_i \cos\omega_j t_i$ and $\sum_{i=1}^{n} h_i \sin\omega_j t_i$ where $j = 1..n/2$. Practical algorithms for the latter were delayed until the introduction of the **Fast Fourier Transform (FFT)** in the mid-sixties, a procedure requiring that n be a power of two (e.g., $n = 1024, 2048, 4096$).

Tidal harmonic analysis at known frequencies is an easier proposition to begin with because, in place of n^2 calculations for the cosine and sine transforms, only $n \cdot m$ multiplication and addition calculations are needed to compute [SXY] with m (the number of tidal constituents) usually varying between 9 and 10 for a 29-day tidal analysis of 696 hourly heights.

MATLAB's highly efficient matrix routines take this process one step further by making the calculations in Eq. (10.2) an almost trivial exercise in matrix algebra – so much so that it is expedient in each analysis to compute both $[SSX]$ and $[SXY]$ directly without regard to orthogonal properties, then invert $[SSX]$ for pre-multiplication with $[SXY]$ to obtain the unknowns matrix $[A]$.

10.4 THE MATLAB ADVANTAGE

A surprisingly short and highly efficient MATLAB program justifies the above claim. We start with an $n \times 1$ vector X containing a column of ones – a vector soon to become a matrix as other columns are successively added in pairs. Given an $n \times 1$ data vector Y with linear trend removed, a $1 \times n$ time vector t, and an $m \times 1$ tidal constituent frequency vector w, only five lines of MATLAB M-file code are needed to construct and solve the harmonic version of Eq. (10.2):

```
X = ones(n,1);
  for j = 1:m
    X = [X  cos(w(j)*t)'  sin(w(j)*t)'];
  end
A = (X'*X)\(X'*Y);
```

The backslash operator (\ ) in the last equation accomplishes left matrix division, the equivalent of left matrix inversion and pre-multiplication with the right matrix to obtain A, the unknowns matrix.

Virtually hidden is another remarkable feature that is especially useful in the analysis of tidal records with 'missing' data. As in ordinary regression analysis, the data are treated as x-y pairs ordered in time, but not necessarily in equal time increments throughout the series. An occasional gap, especially in data gathered at a high sampling rate, can usually be ignored (the mean value is most sensitive here). However, the sampling rate itself cannot be ignored without the risk of data aliasing and the introduction of variance at incorrect frequencies. Under-sampling a high frequency process yields a useless record in advance of any analysis.

10.5 SERIAL DATE AND TIME - MICROSOFT® EXCEL

The tidal analysis and prediction programs I've written in MATLAB utilize the date and time formats contained in the Microsoft Excel workbook. This well-known spreadsheet software specifies January 1, 1900, as serial day 1 (serial hour 24) with the time origin at midnight ending December 30, 1899. The specification is transparent to the user who views dates and times in their customary format in columns of the Excel worksheet. All of the tidal calculations in the programs presented here use the EXCEL time origin, although MATLAB itself recognizes a different origin in its own date and time routines. Following the EXCEL protocol, the elements of the vector t, as shown in the above MATLAB programming example, are serial times in hours and fractions of an hour from the EXCEL time origin. Local standard time (LST) is used throughout with no adjustment for daylight savings time (that adjustment can be made at a later point; i.e., when making predictions).

The advantages of using a single time origin throughout become apparent when making tidal predictions. Traditionally, tidal predictions are 'annual events'. A new **tide table** is issued every year, like a calendar, with the time origin for the predictions set at the beginning of the year. Since the origin differs from year to year, continual modification of the original coefficients $A_0, A_1, B_1, ..., A_m, B_m$ obtained from tidal analysis is required. This modification, which requires cosine argument adjustments involving phase based on the equilibrium tide model is not required if one uses the same time origin throughout (see Chapter 8, Sec. 8.2).

10.6 WATER LEVEL AND WATER CURRRENT ANALYSIS USING HAMELS

A more convenient form of Eq. (10.3) for modeling the height of the tide (or bi-directional water currents) is

$$h(t) = h_0 + \sum_{j=1}^{m} R_j \cos(\omega_j t - \phi_j) \qquad (10.4)$$

where $h_0 = A_0$ is the mean water level (or mean current) and R_j is the amplitude of the jth tidal constituent with frequency ω_j and phase ϕ_j, assuming a total of m constituents in the model. After solving for the unknowns in Eq. (10.3), the amplitude and phase values for Eq. (10.4) are obtained using polar coordinate conversion

$$R_j = \sqrt{A_j^2 + B_j^2}$$
$$\phi_j = \arctan(B_j / A_j)$$

As discussed in Chapter 8, R_j is an amplitude that - for the lunar constituents - varies periodically from year to year according to the 18.6-year cycle of the lunar nodes and is reduced to a mean value, H_j, after division by the corresponding nodal factor f (a factor governed by the mean longitude of the ascending lunar node). Likewise the phase angle ϕ_j for the lunar constituents is adjusted to

$$\kappa_j^* = \phi_j + u_j$$

where u_j is the change in phase of the jth lunar constituent during the nodal cycle. Unless there are significant physical changes in the waterway where a series of tidal observations has been recorded and analyzed, H_j and κ_j^* in theory do not change with time and thus allow tidal predictions to be made in future years. For that reason they are referred to as the *tidal harmonic constants* at the location where the measurements were made.

How 'constant' are the tidal constants? - The term 'constant' does not mean that we will get the same values of H_j and κ_j^* for every new data series analyzed at a particular tide station. This is, after all, a sampling problem in which the constants are derived as sample estimates subject to error. All geophysical data involve *measurement error* that introduces a certain amount of random 'noise' into the sampled data. Except for constituents unusually close to one another in frequency (see Chapter 7, Sec. 7.6), this accounts for most of the variability in H_j and κ_j^* that occurs in, say, monthly analyses conducted at a single station. In addition, there is often a further, contributing error that is weather dependent and not entirely random but event-driven and dependent on the magnitude of the weather event.

Model efficiency - Care should be taken not to confuse the error in determining H_j and κ_j^* with model efficiency in explaining observed variations in water level. Eq. 10.4 represents a model of the *astronomical tide*, a model devised to represent the *regular part* of the signal due to tidal forcing by the moon and sun as opposed to an *irregular part* due to weather-related effects, primarily wind-stress and atmospheric pressure changes that act on the water's surface. In analytical terms, the astronomical part of the signal is regarded as steady, complex periodic, and unbounded in its extent; the non-astronomical part is regarded as non-steady, quasi-periodic or *transient*, lasting only a little longer than the weather event forcing the water level to change. As illustrated by the Persian Gulf examples presented in Chapter 7, it is quite possible to encounter *apparent* errors in least squares data fitting with Eq. 10.4 that reflect primarily the relative magnitude of transient events – which can be large - compared with the fixed amplitudes of the harmonic constituents comprising the local astronomical tide. Even when the least squares fit to the data is low according to the Reduction-in-Variance statistic, the model's efficiency in predicting the astronomical tide may be quite high. The only practical remedy for increasing confidence in the result predicted by Eq. 10.4 is to verify that the residual signal (observed minus predicted) contains no variance at tidal frequencies. As previously stated in Chapter 7, the Reduction-in-Variance statistic

$$RV = \frac{\sum [h(t) - h_0]^2}{\sum [h_t - h_0]^2}$$

expresses the fraction of the total variance (mean square variation in water level or water current) that is accounted for by the specific combination of harmonic constituents used in Eq. 10.4. At this point it might be appropriate to ask whether one could go a step further and determine that part of the total variance accounted for by a single harmonic constituent – M_2 for example. Once again, the difference between harmonic analysis and spectral analysis becomes critically important. If M_2 as well as the other tidal frequencies included in the analysis should happen to be Fourier frequencies, then the total variance (energy or mean square value of the series) could be expressed as

$$\frac{1}{n}\sum_{i=1}^{n}(h_i - h_0)^2 = \frac{1}{2}\sum_{j=1}^{n/2} R_j^{\ 2}$$

if n is an even number. This is the so-called *mean square decomposition* that lies at the heart of spectral analysis and, if applicable, it would allow the fractional variance associated with each constituent to be determined as one-half the square of its amplitude, R_j. But the individual tidal constituents requiring representation in Eq. 10.4 cannot all be multiples of any one fundamental frequency and therefore, as a set, fail to qualify as Fourier frequencies. Consequently, to derive an estimate of the mean square value associated with a tidal constituent, such as M_2, would require two least squares fittings of the data with Eq. 10.4 – one for a tidal model including the constituent and another for the same model excluding it. This is the first step in the analysis of variance procedures for stepwise multiple linear regression models.

10.6 DETERMINING THE PRINCIPAL AXIS FOR WATER CURRENTS

A time series of horizontal water current measurements has a total variance equal to the sum of the variances in the vector components U_i and V_i. Expressing the series $i = 1..n$ as a bivariate data matrix $[U\ V]$ with the bivariate means removed, the variances are easily found as the diagonal elements of the square variance-covariance matrix VCV where [3]

$$VCV = \frac{1}{n}\begin{bmatrix}\sum U^2 & \sum UV \\ \sum UV & \sum V^2\end{bmatrix} = \frac{1}{n}[U\ V]'*[U\ V] \qquad (10.5)$$

Unless we have a very unusual series in which the values comprising one of the components, U or V, are all constant, both elements will contain variance greater than zero and the question is, can we perform some operation on the data matrix that will *maximize* the amount of variance in either U or V? We can - by using a technique known as **Principal Components Analysis** (PCA). PCA transforms the data in just this way after extracting the eigenvectors from the variance-covariance matrix determined from Eq. 10.5. We can do this quite easily in MATLAB using the *eig* function expressed as

$$[EV] = \begin{bmatrix} E_{11} & E_{12} \\ E_{21} & E_{22} \end{bmatrix} = eig(VCV)$$

The term *principal components* simply refers to the column elements in the eigenvector matrix, $[EV]$. The two that are present here (the two columns in the above

---

[3] In using the variance-covariance matrix, we assume the data are multivariate normally distributed.

matrix) are eigenvectors serving as coordinates for two-dimensional data transformation according to the following matrix equation:

$$[U_P \ V_P] = [U \ V] * [EV] \quad (10.6)$$

In practice, $[EV]$ is nothing more than a **rotation matrix** – it rotates the coordinate axes that serve as a reference for multivariate data – to the principal axes. But the rotation has a quite remarkable result; In the present two-dimensional case, the new variable referencing the first principal axis, U_P, will contain the maximum variance that is possible for any axis position while the other, V_P, will contain the least.

To understand how the transformation from $[U \ V]$ to $[U_P \ V_P]$ actually results from axis rotation, see Figure 7.11 in Chapter 7. This figure suggests that surface current data at Richmond, California, are approximately bivariate normally distributed and shows that none of the data points (points representing the tips of current vectors plotted on the U and V coordinate axes) have changed their positions relative to one another as a result of axis rotation. But we see that U_P, the data projected onto the principal axis (axis with reciprocal headings 331°-151°) displays maximum variance while V_P projected on the orthogonal to the principal axis displays minimum variance.

It is important to note that the principal component axis projecting maximum variance is not a line of 'best fit' in the least squares sense; ie., it has nothing to do with correlation or regression.

Appendix 1:

The least squares digital filter

The ideal *low-pass digital filter* does not exist. But if it did, it would smooth a time series by removing all periodic motion oscillating above a specified *cutoff frequency* while retaining oscillations at or below that exact same frequency unmodified. An ideal 36-hour filter, for example, would 'pass' all periodic serial motion at frequencies less than or equal to 1/36 cycles/hour with no attenuation in amplitude and 'stop' all oscillations of higher frequency by reducing their amplitudes to zero. The reality is that amplitude attenuation occurs over a range of frequencies (called the *transition band*) as specified by a *frequency response* or *transfer function*. Thus 1/36 (0.67 cpd) simply marks the 50 per cent response point on the frequency scale within that band (Fig. A1.1). Although we cannot eliminate the transition band and the compromised information that lies within it, we can reduce its width – at a cost. The least squares approach described by Bloomfield (1976) allows a class of linear filter known as a *finite impulse response* (FIR) filter to be systematically designed with that goal in mind.

The general linear filter accomplishes smoothing through a weighted moving average

$$h_t' = \sum_{k=a}^{b} w_k h_{t-k}$$

where $w_a, w_{a+1}, ..., w_b$ is a series of $2m+1$ weights and h_t, h_t' are the input and output series, respectively. If the input series consists of n points, the output series will consist of only $n-2m$ points. If the filter is symmetric ($w_{-k} = w_k$), the transfer function is

$$G(\omega) = \sum_{k=a}^{b} w_k \cos(\omega k) \quad (A1.1)$$

which is the Fourier transform of the filter weights (see Ch. 10, Sec. 10.3). Given a cutoff frequency, ω_c, the least squares principle is used to approximate an ideal filter with response

$$M(\omega) = 1 \quad (0 \leq \omega \leq \omega_c)$$
$$= 0 \quad (\omega_c < \omega \leq \pi) \quad (A1.2)$$

by minimizing the squared difference between it and the design filter (Eq. A1.1) over the range of frequencies $-\pi$ to π. The resulting least squares filter response function is

168 Appendix 1: The least squares digital filter

$$H(\omega) = h_0 + 2\sum_{k=1}^{m} h_k \cos \omega k \quad (A1.3)$$

in which the corresponding filter weights h_k ($k=1:m$) are determined as the Fourier coefficients

$$h_k = \frac{1}{\pi} \int_0^{\omega_c} \cos \omega k \, d\omega = \frac{\sin \omega_c k}{\pi k}$$

and

$$h_0 = \frac{\omega_c}{\pi}$$

After specifying the cutoff frequency, the *width* of the filter and thus the steepness of the response curve at ω_c may be controlled by choosing a suitable number of weights as specified by the parameter m. Finally, to reduce a commonly observed tendency of the response curve to overshoot the filter limits as stated in Eq. A1.2 (Gibbs's phenomenon), each filter weight is multiplied by the following *convergence factor*:

$$\frac{\sin 2\pi k /(2m+1)}{2\pi k /(2m+1)} \quad (A1.4)$$

Fig. A1.1 illustrates the response function for a 36-hour low pass filter obtained with Eqs. A1.3 and A1.4. The filter width fixes the transition band between R10 and R90 as shown below after choosing m = 36. Note that this comes at a cost of 72 data points (72 hours) removed from the ends of a 29-day (696-hour) series of hourly tidal heights.

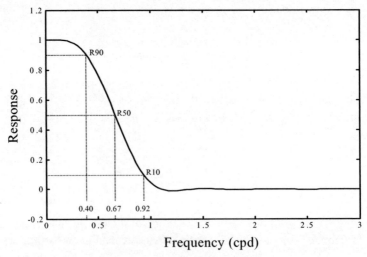

Fig. A1.1. Frequency response diagram for a least squares linear filter with $\omega_c = 0.67\, cpd$ and m = 36. R90=90% response.

Appendix 1: the least squares digital filter

Function LOPASS.M -The following MATLAB function accomplishes least squares filtering of the residual from a water level or water current record after fitting with an astronomical tide model (see Appendices 5 and 6).

```
function xx=lopass(yy,n,omega,ns)
%-------------------------------------------------------------------
% Bloomfield's least squares FIR filter
% yy is the input series (vector of length n)
% omega is the cutoff frequency in radians/unit time
% ns is the filter width index
% lopass returns a filtered series vector of length n-2*ns
%-------------------------------------------------------------------
ns1=ns-1; ns2=2*ns;
lim=n-ns2; ho=omega/pi;
con=2*pi/(ns2+1); sum=ho;
for i=1:ns
  h(i)=ho*(sin(i*omega)/(i*omega))*(sin(i*con)/(i*con));
  sum=sum+2*h(i);
end
ho=ho/sum;
for i=1:ns
  h(i)=h(i)/sum;
end
for i=1:lim
  temp=ho*yy(i+ns);
  for j=1:ns
    temp=temp+(yy(i+ns+j)+yy(i+ns-j))*h(j);
  end
  yy(i)=temp;
end
%compute/plot Fourier transfer function
for j=1:n/2
  f(j)=j/n;
  w(j)=2*pi*f(j);
  H(j)=ho;
  for i=1:ns
    H(j)=H(j)+2*h(i)*cos(i*w(j));
  end
end
figure
f=f*24; aq=round(length(f)/3);
ag=round(length(H)/3);
plot(f(1:aq),H(1:ag))
ylabel('Response')
xlabel('Frequency (cpd)');
Tc=round(2*pi/omega);
legend(['Tc=',num2str(Tc),'h'])
xx=yy;
```

Appendix 2:

Calculation of lunar node factors f and u

The lunar node factors, f and u, are used in tidal prediction formulas as described in Chapter 8 (Sec. 8.1, Eq. 8.1; Sec. 8.2, Eq. 8.4). They account for small variations in the amplitude and phase of the lunar tidal constituents that occur from year to year over the 18.61-year cycle of the lunar nodes (Ch. 2, Sec. 2.8). These variations differ among the constituents involved but all are a consequence of the moon's oscillation about the plane of the ecliptic. This motion is a function of the mean longitude of the ascending lunar node, an *orbital element* with the symbol N which continually changes at a rate of -0.0022 degrees per mean solar hour (the value is negative because the motion of the lunar nodes is westward along the ecliptic). For the zero hour GMT, the formula that determines N in degrees is

$$N = 259.16 - 19.3282(Y - 1900) - 0.0530(D + i)$$

where Y is the year, D is the number of days following January 1 in the year Y and i = the number of leap years between 1900 and Y-1 inclusive, or

$$i = \text{integer } \{(Y-1901)/4\}.$$

The following formulas are used to determine f and u for the nine tidal constituents used in *Simply Tides* and *Simply Currents* (u is given in degrees):

$$f_{M2} = f_{N2} = f_{MS4} = 1.000 - 0.037 \cos N$$
$$u_{M2} = u_{N2} = u_{MS4} = -2.1 \sin N$$

$$f_{K1} = 1.006 + 0.115 \cos N - 0.009 \cos 2N$$
$$u_{K1} = -8.9 \sin N + 0.7 \sin 2N$$

$$f_{O1} = 1.009 + 0.187 \cos N - 0.015 \cos 2N$$
$$u_{O1} = 10.8 \sin N - 1.3 \sin 2N + 0.2 \sin 3N$$

$$f_{M4} = f_{M2}^2 \qquad f_{M6} = f_{M2}^3$$
$$u_{M4} = 2 u_{M2} \qquad u_{M6} = 3 u_{M2}$$

The solar constituents S_2 and S_4 and the seasonal constituents Sa and Ssa are not affected by the lunar node cycle ($f=1$, $u=0$).

Appendix 3:

Tide analysis and prediction: *Simply Tides*

Simply Tides is a set of MATLAB® programs designed for analyzing a short series of water level observations (one to six months) and for making tidal predictions with the tidal constants obtained from multiple 29-day harmonic analyses. To ensure ease of use, each program is accessed through a Graphical User Interface (GUI) as shown below. A GUI is your entry to object-oriented tasking where you make choices as needed by pointing and clicking with your mouse rather than following a script with branching commands and loops (linear programming). You have only one command to enter at the MATLAB command line prompt: 'simplytides'. The GUI takes it from there.

Although the programs and procedures are kept simple, the task of analyzing and predicting tides is not automatic. Both tidal and non-tidal components exist and the user should examine these carefully (see Chapter 7, Sec. 7.3). Two windows are employed. The first window shown below, *Simply Tides: Analysis*, allows quick selection of an Excel data file by double-clicking its .xls file name in a list box. The file is then analyzed and a 29-day plot is displayed. After examining the OPR plot (Observed, Predicted, Residual - see Fig. 7.3), the user should select as many three-day plots as needed to: (1) check for data gaps and errors, (2) understand the information shown in the residual curve. You should not accept the results without these steps.

Simply Tides: Analysis

	First_Day	Last_Day	Plot_Day		Water Levels in:	C:\SimplyTides
Plotting Range	1	27	1	PLOT	⊙ meters ○ feet	cbbt20020731t.xls cbbt20020831t.xls cbbt20020930t.xls glpt20021001t.xls glpt20021101t.xls glpt20021201t.xls **glpt20030101t.xls** liverpool200303.xls

Double Click File to Analyze: 31-Dec-2002 19:00:00 0

	Ssa	Sa		Datums		Statistics	
Amplitude	0.000	0.000	MSL	0.000	RMS Error	0.129	
Phase	0.00	0.00	MLLW	0.000	%R_Var	82.12	

○ Enable print to file
○ Residual periodogram

	M2	S2	N2	K1	O1	M4	M6	S4	MS4
Amplitude	0.3372	0.0551	0.0556	0.0677	0.0389	0.0037	0.0052	0.0086	0.0066
Phase	214.5	287.8	120.8	143.0	105.5	81.7	211.9	84.7	169.7

To save harmonic constants
Enter a Station Name (no spaces), then click SAVE

SAVE   ○ Vector-averaged files

Form Number 0.29

172 Appendix 3: Tide analysis and prediction: *Simply Tides*

The amplitude and phase of nine tidal constituents are displayed in the *Simply Tides: Analysis* window after selecting a file for analysis. If the solar semi-annual (Ssa) and solar annual (Sa) tidal constituents are available for the station in question, their amplitude and phase may be entered manually in the appropriate boxes where the numerals 0.000 and 0.00 appear in blue (blue indicates a data entry field). Likewise, the accepted value of the MSL and MLLW tidal datums may be entered in the adjoining boxes for tide stations maintained by the US NOS. Entering a file name in the lower-left box and pressing SAVE will save the resulting set of tidal constants as a MATLAB MAT-file. At stations with more than one 29-day record, tidal constants from subsequent files may be vector-averaged by depressing the radio button marked 'Vector-averaged files' before double-clicking each file in succession and repeating steps (1) and (2) above. Ideally this is done with a continuous sequence of up to six monthly files. Other options include enabling a 'print to file' of the OPR data and generating a plot of the residual periodogram to examine the non-tidal spectrum. Also included in window boxes are RMS error, percent reduction in variance achieved by the tide model, and tidal form number.

A second GUI, *Simply Tides: Prediction*, is available to generate tidal predictions using a tidal constants file generated by *Simply Tides: Analysis*. The list box in the prediction window will display only MAT-files with extension .mat in the current directory (e.g., C:\Simply tides). After double-clicking a file and choosing a month and year, pressing PREDICT will populate the calendar shown below. Pressing a number will generate a plot for that day. Depressing the 'Enable print to file' button generates a text file of 12-minute heights predicted for the month. These may be in feet or meters.

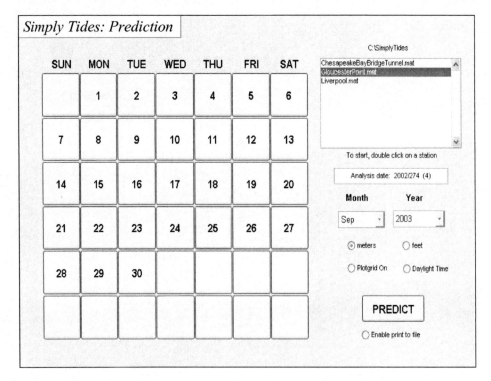

The figure at right is an example of a single daily tide plot produced by *Simply Tides: Predictions*. Additional information is available to the user through *Program Help* windows that pull down from a standard tool bar at the top of the MATLAB GUI (tool bars not shown).

You can download a free copy of *Simply Tides* with some sample Excel data files on the web at MATLAB Central (http://www.mathworks.com/matlabcentral/) and the MATLAB Central File Exchange, 'Companion Software for Books>Earth Sciences' category under the title of this book.

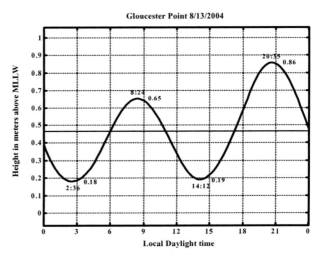

Why 29 days? – A series of 29 days represents a *synodic month*, an interval in which two consecutive conjunctions of lunar phase occurs (spring-neap cycle of 29.53 days) rounded downward to the nearest whole day. It also approximates other lunar cycles in distance and declination whose periods fall between 27 and 28 days. The rationale in applying least squares harmonic analysis is to avoid analyzing fractions of important cycles; thus, 29 days have traditionally been used as a standard for short series tidal analysis in the United States[1]. For many projects in which a tide station is newly established and perhaps temporary, useful analyses and tidal predictions may be made with a set of between one and six months data using the vector-averaging feature of *Simply Tides*. Programs with an extended set of tidal constituents (including the constituent K_2 for example) are not appropriate for an analysis of less than six months[2].

A 29-day least squares harmonic analysis is also useful for another reason: it is arguably the optimum tool for separating *storm surge* from *storm tide* in water level records of any length on the premise that a *synodic* month is the optimum period of time for determining the 'normal' water level and 'normal' tidal behavior that prevails as a large storm arrives and produces a surge. Using less than 29 days, the storm surge contribution clearly begins to weigh more heavily on mean water level and tidal constant estimates. The least squares procedure provides both a reasonable water level estimate as well as a *time-local* estimate of constituent amplitude and phase values reflecting the apparent behaviour of the astronomical tide for that place and time. This is sometimes called *local analysis* as opposed to standard analysis that produces tidal predictions for indefinite periods of time beyond the period of analysis. Put another way, local analysis removes the maximum amount of variance possible (in the least squares sense) from that present in any given water level record at tidal frequencies.

---

[1] A series consisting of 369 days of hourly heights is the US standard.
[2] A series of at least six months is required to separate K_2 from S_2.

Appendix 4:

Current analysis, prediction: *Simply Currents*

Simply Currents is a set of MATLAB® programs designed for analyzing and predicting water currents. They are very similar to the tidal analysis and prediction programs described in Appendix 3. However, because of the difficulty of obtaining a continuous series of water current observations over long periods of time, *Simply Currents: Predictions* permits the user to analyze 14-day as well as 29-day records present in the current directory. Entering 'simplycurrents' as one word at the MATLAB command line prompt activates the Graphical User Interface (GUI). A program help menu describes the data format, which may include either speed and direction or U (east-west) and V (north-south) flow components.

The first GUI window, *Simply Currents: Analysis*, allows quick selection of an Excel data file by double-clicking its file name in a list box. The file is then opened and a UV scatter plot is displayed (see Chapter 7, Sec. 7.5). After examining the 2-D plot, the user will normally enter one of the displayed principal axis headings in the New Axis box under Flood Dir. The choice depends on the user's local knowledge of the approximate direction of flood current (usually the inland direction in a coastal embayment). Otherwise any direction may be entered for whatever reason. The reciprocal heading will be displayed in the box on the right once the Lock Headings button is depressed. Next the ANALYZE button is pressed and the New Axis current is analyzed.

Simply Currents: Analysis

C:\SimplyCurrents

	First_Day	Last_Day	Plot_Day		Current Speed in:	● meters/sec ○ knots
Plotting Range	66	92	66	PLOT		

Files:
CBENC091300m03.5.xls
CBENC091300m18.5.xls
CBENC101300m18.5.xls
COAMA1003m02.7.xls
COAMA1003m14.7.xls
COAMA1003m26.7.xls
MYCB20020320sc.xls
RISB20021101sc.xls

Double Click File to Open
06-Mar-2003 14:45:00 65

○ 14-Day Record

ANALYZE

○ Residual Periodogram
○ Enable print to file

	Flood Dir	Ebb Dir	Axis Variance		Statistics	
Principal Axis	276.4	96.4	Major	0.986	RMS Error	0.068
New Axis	276.0	96.0	Minor	0.014	%R_Var	97.74

● Lock Headings

	M2	S2	N2	K1	O1	M4	M6	S4	MS4
Amplitude	0.5765	0.2634	0.1225	0.0050	0.0140	0.0476	0.0174	0.0048	0.0276
Phase	3.4	105.2	262.3	238.0	103.6	153.1	311.0	202.9	227.1

To save harmonic constants
Enter a Station Name (no spaces), then click SAVE

SAVE   ○ Vector-averaged files

Mean Current   0.01

The above procedure reduces the dimensionally of the water current from two to one; i.e., the UV flow is projected onto the selected axis yielding 1-D flow in the perceived flood and ebb directions about the UV mean. This is done at the expense of a portion of the total variance about this mean but, as in the case of the UV plot at right, a majority of the useful 'information' often lies in the direction of the principal axis. As indicated in the analysis window above, the major axis selected (here the principal axis) accounts for 98.6 percent of the total variance; the minor (orthogonal) axis accounts for only 1.4 percent.

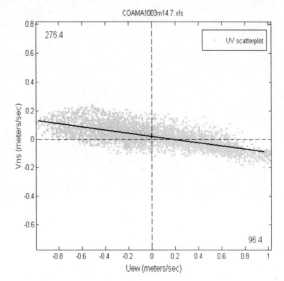

The major axis current is analyzed in the same way as a water level series using the same nine tidal constituents. Note that the reduction-in-variance of the principal axis current by the tidal current model is displayed in the statistics box (97.74 percent reduction in the present example). Unlike a water level series, there is no datum of reference (e.g., MSL or MLLW) other than the mean current determined during the 29-day (14-day) period of the analysis. This number appears in blue in the lower right-hand box on the GUI and may be manually replaced -either with a zero or another mean current value. File saving and vector-averaging of the tidal current constituents is done as described for water levels in Appendix 3 once the user is satisfied with OPR plots and other information regarding data quality (see Fig. 7.18, Ch. 7).

A second GUI, *Simply Currents: Prediction*, is available to generate tidal current predictions using a tidal constants file generated by *Simply Currents: Analysis*. An example is shown below. The list box in the prediction window will display only MAT-files with extension .mat in the current directory (e.g., C:\Simply Currents). After double-clicking a file and choosing a month and year, pressing PREDICT will populate the calendar array. Pressing a number in the array will generate a plot for that day. Depressing the 'Enable print to file' button before pressing PREDICT will generate a text file of 12-minute current speeds predicted for the month. These may be in knots or meters per second; their serial times may reference either Local Standard Time or Local Daylight Time.

Current Plots – One of the reasons for projecting currents onto a principal axis is that it enables a plot of a current curve to be drawn. The alternative is to analyze and determine harmonic constants for both the major and minor axes of a current ellipse and generate dual predictions in a 2-D vector plot. Such plots are more complex and difficult to visualize over time. Except where currents are of the rotary type with an ellipse expanding toward a circle, current plots featuring a sinusoidal curve are arguably more useful – better highlighting diurnal inequalities in current, for example.

Appendix 4: Current analysis, prediction: *Simply Currents*

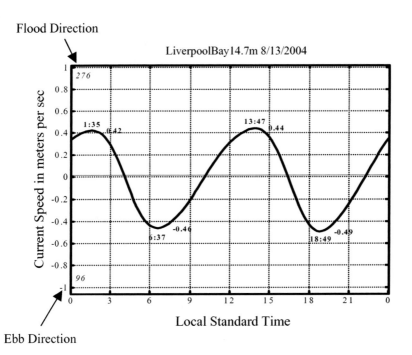

Appendix 5:

Tidal analysis and tidal prediction programs (command line versions)

A5.1. HAMELS_TIDE
Program HAMELS_TIDE (Harmonic Analysis Method of Least Squares: Tides) is a MATLAB M-file designed to analyze 29-day records of tidal heights using MATLAB's 'xlsread' function to read input data (date, time, height) from a Microsoft EXCEL workbook. A total of nine harmonic constituents are included in a standard analysis; other constituents may be added or substituted as needed by modifying the M-file listed below. This is a command line program – not a GUI; after entering the M-file name (HAMELS_TIDE), the user is prompted for an input filename and given options for conducting the analysis. MATLAB function LOPASS.M is required (see Appendix 1).

Standard input format – The recommended source for tidal height data is the online archive for NOAA/NOS water level observations at http://co-ops.nos.noaa.gov. When conducting an NOS search, the user is advised to request historical records that include thirty days or more of hourly heights, choosing 'yyyy/mm/dd hh:mm' as the output format and Local Standard Time (LST) as the time zone. The resulting screen text is transferred to an EXCEL workbook using Windows 'cut' and 'paste' commands.

- Station header information, including data descriptors and related text, should be pasted on the second worksheet in the EXCEL workbook.

- Paste date, time, UV or SD input data as numeric text on the first worksheet.

- The EXCEL data command 'text-to-columns' is used to parse the numeric text on the first worksheet into individual data columns. The file should be saved as an Excel workbook using a file name that identifies the station and date.

- No column or row in the NOS data format, including a leading row of column labels, requires deleting. MATLAB reads the entire array provided there are an equal number of columns in each row. The program selects the data columns used in the analysis.

- The program code contains a vector variable, *vdc,* specifying *valid data columns*; i.e., columns containing date, time, and height of tide in the standard NOS array format. Should a variation in the *vdc* settings be required for this format, the user can reset the elements in the *vdc* vector to select valid column numbers (e.g., *vdc*=[2 4 5]) for the new data array based on order of appearance, left to right.

Appendix 5: Tidal analysis and tidal prediction programs 179

Note: HAMELS_TIDE expects date and time in the EXCEL format. In the EXCEL worksheet the user will see, for example, '9/1/2002' for date and '1:00:00 AM' for time. HAMELS_TIDE automatically works with their numeric equivalents; i.e., 37500 and 0.04167 for the serial date and time in days and fractions beginning January 1, 1900 as day 1. For simplicity, graphical results are displayed in Julian days 1-365.

HAMELS_TIDE output - When analysis is completed, the resulting harmonic constants are automatically saved in an EXCEL workbook named 'hamels_tide_out.xls'. Should the user wish to permanently save this output, the file should be renamed immediately to avoid its being overwritten during a subsequent run. Within this file, zeroes appear in number fields for the amplitude and phase of the seasonal constituents *Sa* and *Ssa*, as well as the fields for the *MLLW* and *MSL* tidal datums. If the data for these fields are available from the NOS archives, they should be used to replace the default (zero) values.

A5.2. ASTRO_TIDE

Program ASTRO_TIDE (Astronomical predictions: tides) uses the harmonic constants derived for a given tide station by its companion program, HAMELS_TIDE, to generate astronomical tide predictions in Local Standard Time (LST) for any given year at that station.

- Assuming harmonic constants for the seasonal constituents *Sa* and *Ssa* have been added to the list of constituents (in the EXCEL workbook file originally named 'hamels_tide_out.xls'), tidal predictions will include the seasonal tide.

- If values for the *MLLW* and *MSL* datum elevations are included, predicted tidal heights will reference *MLLW*. Otherwise, if the datum elevations are both zero (no offset between *MLLW* and *MSL*), predicted tidal heights will reference *MSL*.

Please note that HAMELS_TIDE, ASTRO_TIDE and Microsoft EXCEL all share the same origin for serial date and time: midnight ending December 30, 1899. This enables ASTRO_TIDE to generate tidal predictions for any year after this date without need of the so-called Greenwich equilibrium arguments. This means that the phase lags (kappa values) given in NOS listings of tidal harmonic constants <u>will not work</u> in program ASTRO_TIDE. The seasonal constituents *Sa* and *Ssa* are the sole exception; ASTRO_TIDE utilizes the NOS harmonic constants for these constituents after applying the required Greenwich equilibrium arguments to their phase lags assuming Local Standard Time (LST) is in use.

ASTRO_TIDE Output - ASTRO_TIDE gives the option of generating a tide graph for one day, one month, or one year. The time series of tidal heights shown in each graph are – at the user's option - directed to an EXCEL output file named 'astro_tide_out.xls'. Note that dates and times for the heights listed in this file are serial EXCEL dates – not serial MATLAB dates which use a different time origin. Conversions (transparent to the user) are made between the two where necessary within the MATLAB program.

DISCLAIMER – Tidal predictions made with programs HAMELS_TIDE and ASTRO_TIDE are not intended for operational use in any activity involving vessel

180 Appendix 5: Tidal analysis and tidal prediction programs

navigation and marine safety. The author assumes no responsibility if the programs are used for such purposes.

HAMELS_TIDE.m

```
clear
echo on
% Program HAMELS_TIDE        - Harmonic Analysis Method of Least Squares
% analyzes 29-day time series - heights read from Microsoft EXCEL Workbook
% Tidal harmonic constituents - M2 S2 N2 K1 O1 M4 M6 S4 MS4 (9)
% Time origin for analysis    - Midnight ending December 30,1899 LST
% Data series input format    - NOS/NOAA/CO-OPS hourly heights format
% Time display units          - Serial days from start of year(Julian days)
% Version 2.2 w/LSQ filter    - written by J.D. Boon (VIMS) 11-25-2003
echo off
m2=28.9841042*pi/180;       % M2 frequency (radians/mean solar hour)
s2=30.0000000*pi/180;       % S2 frequency (radians/mean solar hour)
n2=28.4397295*pi/180;       % N2 frequency (radians/mean solar hour)
k1=15.0410686*pi/180;       % K1 frequency (radians/mean solar hour)
o1=13.9430356*pi/180;       % O1 frequency (radians/mean solar hour)
m4=2*m2;                    % M4 frequency (radians/mean solar hour)
m6=3*m2;                    % M6 frequency (radians/mean solar hour)
s4=2*s2;                    % S4 frequency (radians/mean solar hour)
ms4=m2+s2;                  % MS4 frequency(radians/mean solar hour)
w=[m2 s2 n2 k1 o1 m4 m6 s4 ms4]; % frequency vector
NC=length(w);               % number of tidal harmonic constituents
Constituents =' M2 S2 N2 K1 O1 M4 M6 S4 MS4';
%---------------------------
% Set valid data columns
%---------------------------
DTH=[2 3 4];  % Date, Time, Height
%-----------------------------------------------------------
% select spreadsheet containing water level data
%-----------------------------------------------------------
file=input('Enter spreadsheet name <e.g., CBBT20021101.xls>: ','s');
Y=xlsread(file); uc=input('Water levels recorded in meters(1) or feet(2): ');
if uc==1 units='meters'; else units='feet'; end
%-----------------------
% get date_time_height
%-----------------------
  vdc=DTH; i=1;
  for j=1:length(Y)
    if Y(j,vdc(1))>0 X(i,[1 2 3])=Y(j,vdc); i=i+1; end
  end
  [n,m]=size(X);
%-----------------------
% initialize data series
```

Appendix 5: Tidal analysis and tidal prediction programs

```
%------------------------
sd=X(:,1)+X(:,2);                    % serial days since 1900 plus fractional day
leap=fix((sd(1)-366)/1460)           % leap years since 1900
ys=fix((sd(1)-leap-1)/365)+1900;     % year starting data series
sds=fix((ys-1900)*365+leap+1);       % serial day starting year
jd=sd-ones(n,1)*sds(1);              % Julian day plus fractional day
jt=jd*24;                            % Julian time in hours
st=sd*24;                            % serial time in hours since 1900
sr=round(1/((jd(2)-jd(1))*24));      % sampling rate (no. readings per hour)
Ne=24*29*sr;                         % series length expected for 29-day analysis
%------------------------
% Check series length
%------------------------
if sd(n)-sd(1)<29
      series_length=' less than 29 days - program terminated'
break; end
%------------------------------
% find first whole Julian day
%------------------------------
c=mod(jd(1),1)>(0.0416/sr);
jdf=fix(jd(1))+c; dte=sd(1)+datenum('30-Dec-1899');
first_record=[datestr(dte),' ',num2str(jd(1))]; jds=jdf;
jdlast=fix(jd(n,1)); jdsec=jdlast-round(Ne/(24*sr));
%---------------------
% start analysis loop
% select starting day
%---------------------
while jds>0
  vnum=0; SRate_Start1_Start2_lastday=[sr jdf jdsec jdlast]
  while vnum==0
    vnum=1;
    jds=input('Enter Julian Day to start a 29-day analysis <0 to quit>: ');
    if jds==0 break; end
    jde=jds+Ne/(24*sr);
    if jds<jdf vnum=0; elseif jde>jdlast vnum=0; end;
  end
  if jds==0 break; end
%---------------------
% find jds,jde indices
%---------------------
  for i=1:n
    if jd(i)>=jds is=i; break; end
  end
  for i=1:n
    if jd(i)>=jde ie=i-1; break; end
  end
  N=ie-is+1;
  records_expected_found=[Ne N]    % number of records expected _ number found
```

Appendix 5: Tidal analysis and tidal prediction programs

```
h=X(is:ie,3);                    % height vector selected for analysis
t=st(is:ie)';                    % serial time vector selected for analysis
ajt=jt(is:ie)';                  % Julian time vector in hours
hbar=mean(h);                    % data mean
td=t-mean(t);                    % time in deviate form
hd=h-mean(h);                    % data in deviate form
trend=td*hd/(td'*td');           % data trend
hdn=hd-td'*trend;                % trend removed
%-----------------------------------------------
% set up sums of squares matrix for N points
% solve normal equations for least squares fit
%-----------------------------------------------
  A=ones(N,1);
    for j=1:NC
     A=[A cos(w(j)*t)' sin(w(j)*t)'];
    end
  a=(A'*A)\A'*hdn;
%-----------------------------------------------
% determine mean, amplitude and phase
% for NC harmonic constituents
%-----------------------------------------------
   for j=1:NC
     R(j)=sqrt(a(2*j)^2+a(2*j+1)^2);
     Z(j)=atan2(a(2*j+1),a(2*j));
    end
   Ro=a(1);
%-------------------------
% predict tidal heights
%-------------------------
Constituents
Pdeg=mod(Z*180/pi,360);
hp=Ro+R*cos(w'*t-Z'*ones(1,N));     % predicted tides for series analyzed
hp=hp+hbar+trend*td; hp=hp';        % predicted tides plus series mean and trend
%-----------------------------
% display LSQ fit statistics
%-----------------------------
    rms=sqrt((h-hp)'*(h-hp)/N);            % RMS error
    Vh=(h-hbar)'*(h-hbar)/(N-1);           % Total variance for series analyzed
    Vhp=(hp-hbar)'*(hp-hbar)/(N-1);        % Variance accounted for by tide predictions
    pv=Vhp*100/Vh;                         % Percent reduction in variance
    fn=(R(4)+R(5))/(R(1)+R(2));            % Tidal form number
    RMSError_RVar_FormNumber=[rms pv fn]   % display stats

%-----------------------------------------------
% Set filter cutoff for residual heights analysis
%-----------------------------------------------
TF=36;                           % FILTER CUTOFF period in hours
jFC=round(n/TF);                 % nearest Fourier frequency number
```

Appendix 5: Tidal analysis and tidal prediction programs

```
cutoff=2*pi*jFC/n;                        % FILTER CUTOFF freq in radians/hr
%-------------------
% Set filter width
%-------------------
ns=36; n=length(hdn);
%-------------------------------------------------------------------------
% call LOPASS routine using cutoff and filter width ns=36
%-------------------------------------------------------------------------
hh=lopass(hdn,n,cutoff,ns);
lim=n-2*ns;
hh=hh(1:lim);
sjds=jds+ns/(24*sr);
sjd=sjds:1/(24*sr):sjds+(lim-1)/(24*sr);
%------------------------------------
% compute orbital elements
% convert R to H (18.6-yr mean)
%------------------------------------
% starting day + leapyear days since 1901
  Di=jds+fix((ys-1901)/4);
% Mean longitude of ascending lunar node
  Nm=259.16-19.3282*(ys-1900)-0.053*Di;
%--------------------------------------
  Nmr=Nm*pi/180; fM2=1-0.037*cos(Nmr);
  fK1=1.006+0.115*cos(Nmr)-0.009*cos(2*Nmr);
  fO1=1.009+0.187*cos(Nmr)-0.015*cos(2*Nmr);
  H(1)=R(1)/fM2; H(2)=R(2); H(3)=R(3)/fM2;
  H(4)=R(4)/fK1; H(5)=R(5)/fO1;
  H(6)=R(6)/(fM2^2); H(7)=R(7)/(fM2^3);
  H(8)=R(8); H(9)=R(9)/fM2; uM2=-2.1*sin(Nmr);
  uK1=-8.9*sin(Nmr)+0.7*sin(2*Nmr);
  uO1=10.8*sin(Nmr)-1.3*sin(2*Nmr)+0.2*sin(3*Nmr);
  Pdeg(1)=Pdeg(1)+uM2; Pdeg(3)=Pdeg(3)+uM2;
  Pdeg(4)=Pdeg(4)+uK1; Pdeg(5)=Pdeg(5)+uO1;
  Pdeg(6)=Pdeg(6)+2*uM2; Pdeg(7)=Pdeg(7)+3*uM2;
  Pdeg(9)=Pdeg(9)+uM2;
%---------------------------------
% display amplitude and phase
% store results on new spreadsheet
%---------------------------------
Amplitude_Phase=[H;Pdeg];                 % amplitude and phase(degrees)
fout=fopen('hamels_tide_out.xls','w');    % opening data outfile
fprintf(fout,'%s',[file(1:4),' ',units]); % writing header info
fprintf(fout,'%s',['  Y',file(5:8)]);
fprintf(fout,'%s\n',num2str(jds)); j=1;
for i=1:NC
   tc=Constituents(j:j+3); j=j+3;         % write results to outfile
   fprintf(fout,'%s',tc);
   fprintf(fout,'%8.4f %8.4f\n',H(i),Pdeg(i));
```

```
       end
       fprintf(fout,'%s\n',' SSA  0.0000 0.0000');   % write solar semiannual amp, phase
       fprintf(fout,'%s\n',' SA   0.0000 0.0000');   % write solar annual amp, phase
       fprintf(fout,'%s\n','TDEL  0.0000 0.0000');   % write MLLW, MSL datum elevations
       fprintf(fout,'%2.0f',uc);                     % write units code: meters(1) feet(2)
       fprintf(fout,'%s\n','  MSL    MLLW');         % NOTE: substitute NOS values above
       status=fclose(fout);
%--------------
% main plot
%--------------
       figure
       jdp=ajt(1:N)/24; hb=hbar*ones(1,N);
       plot(jdp,h,'r-',jdp,hp,'b-',jdp,h-hp,'g-',jdp,h*0,'k:',jdp,hb,'k-')
       title(file)
       xlabel('Julian Day')
       ylabel(['Water level in ',units])
       hold on
       plot(sjd,hh,'k-');
       Legend('Observed','Predicted','Residual')
       hold off
%--------------------
% three-day plots
%--------------------
       p=1; jmax=jde-3;
       while p>0
         vnum=0;
         while vnum==0
           p=input('enter Julian day for three-day plot <0 to exit>: ');
           vnum=1; if p==0 break; end
           if p<jds vnum=0; elseif p>jmax vnum=0; end
         end
         if p==0 break; end
         j=(p-jds)*24*sr+1; k=72*sr-1; jt3=ajt(j:j+k)/24;
         ho3=h(j:j+k); hp3=hp(j:j+k);
         figure
         plot(jt3,ho3,'r-',jt3,hp3,'b-',jt3,ho3-hp3,'g-',jt3,ho3*0,'k--')
         title(file)
         xlabel('Julian Day')
         ylabel(units)
       end
end
Harmonic_Constants_file=['hamels_tide_out.xls']
```

ASTRO_TIDE.m

```
clear
echo on
% Program ASTRO_TIDE        - Astronomical Tide Prediction, command version
```

Appendix 5: Tidal analysis and tidal prediction programs

```
% generates predicted tides          - tidal height in file named 'astro_tide_out.xls'
% Time origin for prediction         - Midnight ending December 30,1899 LST
% Tidal harmonic constituents        - NC constituents from program HAMELS_TIDE
% Numeric Input (X matrix)           - NC rows of amplitude(m or ft) and phase(deg)
% Numeric Input (X matrix)           - one row of solar semiannual(Ssa) a&p (optional)
% Numeric Input (X matrix)           - one row of solar annual(Sa) amp&phse (optional)
% Numeric Input (X matrix)           - one row of MLLWand MSL elevations (optional)
% Version 2.1 for EXCEL              - written by J.D. Boon (VIMS) 12/5/2003
echo off
m2=28.9841042*pi/180;          % M2 frequency (deg/mean solar hour)
s2=30.0000000*pi/180;          % S2 frequency (deg/mean solar hour)
n2=28.4397295*pi/180;          % N2 frequency (deg/mean solar hour)
k1=15.0410686*pi/180;          % K1 frequency (deg/mean solar hour)
o1=13.9430356*pi/180;          % O1 frequency (deg/mean solar hour)
m4=2*m2;                       % M4 frequency (deg/mean solar hour)
m6=3*m2;                       % M6 frequency (deg/mean solar hour)
s4=2*s2;                       % S4 frequency (deg/mean solar hour)
ms4=m2+s2;                     % MS4 frequency(deg/mean solar hour)
sa= 0.0410686*pi/180;          % Sa  frequency(deg/mean solar hour)
ssa=0.0821373*pi/180;          % Ssa frequency(deg/mean solar hour)
w=[m2 s2 n2 k1 o1 m4 m6 s4 ms4 ssa sa]; % frequency vector
%----------------------------------
% Enter harmonic constants file
%----------------------------------
file=input('Enter filename <e.g., ChesBayBrTunnel.xls>: ','s');
[X,A]=xlsread(file); [n,m]=size(X);
H=X(1:n-2,2); Pdeg=X(1:n-2,3);
R=H; P=Pdeg;
%--------------------------------------
% Enter prediction date and units
%--------------------------------------
year=input('  Enter year of predictions <yyyy>: ','s');
  nu=input('  heights in meters(1) or feet(2) : ');
tpred=input('   plot one day(1),month(2),year(3): ');
%--------------------------------------
% compute orbital elements
% convert H (18.6-yr mean) to R
%--------------------------------------
% starting year; leap-yr days since 1900+183
  ys=str2num(year); Di=fix((ys-1901)/4)+183;
% Mean longitude of ascending lunar node
  Nm=259.16-19.3282*(ys-1900)-0.053*Di;
%-----------------------------------------------------
  Nmr=Nm*pi/180; fM2=1-0.037*cos(Nmr);
  fK1=1.006+0.115*cos(Nmr)-0.009*cos(2*Nmr);
  fO1=1.009+0.187*cos(Nmr)-0.015*cos(2*Nmr);
  R(1)=H(1)*fM2; R(3)=H(3)*fM2; R(4)=H(4)*fK1;
  R(5)=H(5)*fO1; R(6)=H(6)*(fM2^2);
```

```
   R(7)=H(7)*(fM2^3); R(9)=H(9)*fM2;
   uM2=-2.1*sin(Nmr); uK1=-8.9*sin(Nmr)+0.7*sin(2*Nmr);
   uO1=10.8*sin(Nmr)-1.3*sin(2*Nmr)+0.2*sin(3*Nmr);
   P(1)=Pdeg(1)-uM2; P(3)=Pdeg(3)-uM2;
   P(4)=Pdeg(4)-uK1; P(5)=Pdeg(5)-uO1;
   P(6)=Pdeg(6)-2*uM2; P(7)=Pdeg(7)-3*uM2; P(9)=Pdeg(9)-uM2;
   P(11)=Pdeg(11)-280;  % Padj for Greenwich equilibrium phase, Sa(2000)
   P(10)=Pdeg(10)-200;  % Padj for Greenwich equilibrium phase, Ssa(2000)
   P=P*pi/180; MSL=X(12,2); MLLW=X(12,3); delh=MSL-MLLW;
%-----------
% Set units
%-----------
eu=X(13,1); cf=1;
if nu>1
   units='Feet';
   if eu<2 cf=1/0.3048; end
else
   units='Meters';
   if eu>1 cf=0.3048; end
end
bd=['01/01/',year]; bdv=datevec(bd); spc=' ';
%-------------------------------
% Generate, plot predictions
%-------------------------------
if tpred==1
   ss(1:19)=spc; sdat=input([ss,'month/day <mm/dd>: '],'s');
   sdv=datevec(sdat); sdv(1)=str2num(year);
   sd=datenum(sdv)-datenum('30-Dec-1899'); ed=sd+1;
   ti=0.5; t=[sd*24:ti:ed*24-ti]; N=length(t); tp=t/24;
   hp=delh*ones(1,N)+R'*cos(w'*t-P*ones(1,N)); hp=hp*cf;
     as=R(10:11); ws=w(10:11); ps=P(10:11);
   hps=delh*ones(1,N)+as'*cos(ws'*t-ps*ones(1,N)); hps=hps*cf;
   stamp=[file(1:4),' ',year,' ',sdat];
   figure
    plot(tp,hp,'b-',tp,hps,'r-')
    datetick('x',16)
    title(stamp)
    xlabel('Local Standard Time (hours)')
    if delh==0
       ylabel([units,' Above MSL'])
       datum='MSL';
    else
       ylabel([units,' Above MLLW'])
       datum='MLLW';
    end
    grid on
elseif tpred==2
   ss(1:23)=spc; ms=input([ss,'month <1..12>: ']); me=ms+1;
```

```
    bdv(2)=ms; sd=datenum(bdv)-datenum('30-Dec-1899');
    bdv(2)=me; ed=datenum(bdv)-datenum('30-Dec-1899');
    ti=1; t=[sd*24:ti:ed*24-ti]; N=length(t); tp=t/24;
    hp=delh*ones(1,N)+R'*cos(w'*t-P*ones(1,N)); hp=hp*cf;
     as=R(10:11); ws=w(10:11); ps=P(10:11);
    hps=delh*ones(1,N)+as'*cos(ws'*t-ps*ones(1,N)); hps=hps*cf;
    stamp=[file(1:4),' ',year];
    figure
    plot(tp,hp,'b-',tp,hps,'r-')
    datetick('x',6)
    title(stamp)
    xlabel('Local Standard Time (days)')
    if delh==0
       ylabel([units,' Above MSL'])
       datum='MSL';
    else
       ylabel([units,' Above MLLW'])
       datum='MLLW';
    end
elseif tpred==3
    ms=1; me=ms+12;
    bdv(2)=ms; sd=datenum(bdv)-datenum('30-Dec-1899');
    bdv(2)=me; ed=datenum(bdv)-datenum('30-Dec-1899');
    ti=1; t=[sd*24:ti:ed*24-ti]; N=length(t); tp=t/24;
    hp=delh*ones(1,N)+R'*cos(w'*t-P*ones(1,N)); hp=hp*cf;
     as=R(10:11); ws=w(10:11); ps=P(10:11);
    hps=delh*ones(1,N)+as'*cos(ws'*t-ps*ones(1,N)); hps=hps*cf;
    stamp=[file(1:4),' ',year]; tx=tp-(sd-1)*ones(1,N);
    figure
    plot(tx,hp,'b-',tx,hps,'r-')
    V=axis; V(1)=1; V(2)=366;
    axis(V)
    if mod(str2num(year),4)==0
       set(gca,'xtick',[1 30 60 90 120 150 180 210 240 270 300 330 366])
    else
       set(gca,'xtick',[1 30 60 90 120 150 180 210 240 270 300 330 365])
    end
    title(stamp)
    xlabel('Julian Days')
    if delh==0
       ylabel([units,' Above MSL'])
       datum='MSL';
    else
       ylabel([units,' Above MLLW'])
       datum='MLLW';
    end
end
sav=input('Save this data? <y/n>: ','s');
```

Appendix 5: Tidal analysis and tidal prediction programs

```
if sav=='y'
   saved_file=['astro_tide_out.xls']
   fout=fopen('astro_tide_out.xls','w');      % opening data outfile
   fprintf(fout,'%s',file(1:4));              % writing header info
   fprintf(fout,'%s',[year,'    ',units]);
   fprintf(fout,'%s\n',['    ',datum]);
   data=[tp' hp'];
   fprintf(fout,'%10.5f %9.3f\n',data');      % writing data to file
   status=fclose(fout);
end
```

Appendix 6:

Current analysis and current prediction programs (command line versions)

A6.1. HAMELS_CURRENT
Program HAMELS_CURRENT (Harmonic Analysis Method of Least Squares: Current) is a MATLAB m-file designed to analyze 29-day (optionally 14-day) records of water current recorded either as east-west, north-south (UV) flow components, or as flow speed and direction (SD) relative to True North. The MATLAB 'xlsread' function is used to read input data (date, time, UV/SD) from a Microsoft EXCEL workbook. A total of nine harmonic constituents are included in a standard analysis; other constituents may be added or substituted as needed by modifying the M-file presented below. After executing the program, the user is prompted for an input filename and is then given options for conducting the analysis, including the choice of *principal axis heading*[1]. The positive flood (U+) current is defined by this heading.

Standard input format – The recommended source for tidal current data is the online archive for NOAA/NOS water current observations found under the 'Infohub' page at http://co-ops.nos.noaa.gov. When using the NOS search facility, the user is advised to request historical records in Local Standard Time (LST) that include one month of 6-minute current speed and direction. The resulting screen output is transferred as text to an EXCEL workbook using Windows 'cut' and 'paste' commands.

- Station header information, including data descriptors and related text, should be pasted on the second worksheet in the EXCEL workbook.

- Paste date, time and UV or SD input data as numeric text on the first worksheet.

- The EXCEL data command 'text-to-columns' is used to parse the numeric text on the first worksheet into individual data columns. The data are then ready for input into MATLAB.

- No column or row, including a leading row of column labels, requires deleting. MATLAB reads the entire array provided there are an equal number of columns in each row. The program selects the data columns used in the analysis.

---

[1] Reciprocal headings are given for this axis – the user must select the one matching the flood direction.

- The program code contains a vector variable, *vdc,* specifying *valid data columns*; i.e., columns containing *date, time, speed,* and *direction* of current in the standard NOS array format. A second format for data selection including date, time, and *orthogonal (U, V) current components* is programmed as an alternate choice. Should a variation in the *vdc* settings be required for either format, the user can reset the elements in the *vdc* vector to select valid column numbers (e.g., *vdc=* [2 4 5 6]) for the new data array based on order of appearance, left to right.

NOTE: HAMELS_CURRENT expects date and time in the EXCEL format. In the EXCEL worksheet the user may see, for example, '9/1/2002 1:00:00 AM' for date and time. HAMELS_UVx automatically works with the numeric equivalent, '37500.04167' for the serial date and time in days and fractions of a day since December 30, 1899. For simplicity, intermediate program results are displayed in Julian days. For example, the program will determine and display a range of valid starting days labeled as 'Start1' and 'Start2'. The user can select any day within this range to begin an analysis.

HAMELS_CURRENT Output – When analysis is completed, the resulting harmonic constants for the principal axis current are automatically saved in an EXCEL workbook named 'hamels_current_out.xls'. Should the user wish to permanently save this output, the file should be renamed immediately to avoid its being overwritten during a subsequent run. The mean speed and direction of the principal axis current is entered in the last row of this file. Reset the mean speed to zero if you do not want the mean current included in subsequent current predictions with program ASTRO_CURRENT.

A6.2. ASTRO_CURRENT

Program ASTRO_CURRENT (Astronomical predictions: tidal current) uses the harmonic constants derived for a given current station by its companion program, HAMELS_CURRENT. Program ASTRO_CURRENT generates astronomical current predictions in Local Standard Time for any given year at that station.

- Although predictions are based on analysis of two-dimensional (UV or SD) current records, ASTRO_CURRENT generates predictions of the one-dimensional alternating current in the directions of maximum variance as defined by the principal current axis and its reciprocal headings.

- Positive (U+) currents are flows in the flood direction as identified by the user during harmonic analysis at the station in question. Negative (U-) currents represent flows in the ebb direction.

Note: HAMELS_CURRENT, ASTRO_CURRENT and Microsoft EXCEL all share the same origin for serial date and time: midnight ending December 30, 1899. This enables ASTRO_CURRENT to generate tidal current predictions for any year after this date without need of the so-called Greenwich equilibrium arguments (see Chapter 8, Sec. 8.2). This means that the phase lags (kappa values) given in NOS listings of the tidal harmonic constants will not work in program ASTRO_CURRENT.

Appendix 6: Current analysis and current prediction programs 191

ASTRO_CURRENT Output: ASTRO_CURRENT gives the option of generating a tide graph for one day, one month, or one year. At the user's option, the time series of tidal heights shown in the current graph are directed to an EXCEL output file named 'astro_current_out.xls'. Dates and times for the heights listed in this file are serial EXCEL dates – not serial MATLAB dates which use a different time origin. Conversions (transparent to the user) are made between the two where necessary within the MATLAB program.

DISCLAIMER – Tidal current predictions made with HAMELS_CURRENT and ASTRO_CURRENT are not intended for operational use in any activity involving vessel navigation and marine safety. The author assumes no responsibility for such use.

HAMELS_CURRENT.m

```
clear
echo on
% Program HAMELS_CURRENT     - Harmonic Analysis Method of Least Squares
% analyzes 29-day (14-day) series  - Excel workbook of UV or SD current data
% tidal harmonic constituents  - M2 S2 N2 K1 O1 M4 M6 S4 MS4 (9)
% time origin for analysis    - Midnight ending December 30,1899 LST
% 1. Speed&Direction format   - NOAA/NOS/PORTS 6-min currents
% 2. UV current input format  - BODC/POL measured current archive
% Version 2.1 for EXCEL input - written by J.D. Boon (VIMS) 12-5-2003
echo off
m2=28.9841042*pi/180;        % M2 frequency (radians/mean solar hour)
s2=30.0000000*pi/180;        % S2 frequency (radians/mean solar hour)
n2=28.4397295*pi/180;        % N2 frequency (radians/mean solar hour)
k1=15.0410686*pi/180;        % K1 frequency (radians/mean solar hour)
o1=13.9430356*pi/180;        % O1 frequency (radians/mean solar hour)
m4=2*m2;                     % M4 frequency (radians/mean solar hour)
m6=3*m2;                     % M6 frequency (radians/mean solar hour)
s4=2*s2;                     % S4 frequency (radians/mean solar hour)
ms4=m2+s2;                   % MS4 frequency(radians/mean solar hour)
w=[m2 s2 n2 k1 o1 m4 m6 s4 ms4];  % frequency vector
NC=length(w);                % number of tidal harmonic constituents
Constituents=' M2 S2 N2 K1 O1 M4 M6 S4 MS4';
%---------------------------
% Set valid data columns
%---------------------------
DTSD=[1 2 4 5];              % Date,Time,Speed,Direction
DUV=[2 3 4];                 % Date&Time,Ue,Vn

%------------------------------------------------------------
% select spreadsheet containing water current data
% select series length to be analyzed (14 or 29 days)
%------------------------------------------------------------
file=input('Enter spreadsheet name <e.g., CBENC091300m19.xls>: ','s');
```

Appendix 6: Current analysis and current prediction programs

```
sera=input('Enter series length to be analyzed (14 or 29 days): ');
Y=xlsread(file);
 spd=input('Current S&D vectors(1) or UV components(2): ');
 uc=input(' recorded in knots(1) or meters/second(2): ');
if uc==1
  uns='Up(kts)'; vns='Vp(kts)'; us='U(kts)'; vs='V(kts)'; units='knots'
else
  uns='Up(mps)'; vns='Vp(mps)'; us='U(mps)'; vs='V(mps)'; units='mps '
end
%--------------------
% get current data
%--------------------
if spd==1
% date_time_spd_dir format
  vdc=DTSD; i=1;
  for j=1:length(Y)
    if Y(j,vdc(1))>0
      X(i,[1 2 3 4])=Y(j,vdc); i=i+1;
    end
  end
 sd=X(:,1)+X(:,2);
elseif spd==2
% date&time_Ue_Vn format
  vdc=DUV; i=1;
  for j=1:length(Y)
    if Y(j,vdc(1))>0
      X(i,[1 2 3])=Y(j,vdc); i=i+1;
    end
  end
  sd=X(:,1);
end
[n,m]=size(X);
%-----------------------
% Check series length
%-----------------------
if sera>14
   if sd(n)-sd(1)<29
      series_length=' 28 days or less - program terminated'
      break; end
elseif sd(n)-sd(1)<14
      series_length=' 13 days or less - program terminated'
      break; end
%--------------------------
% initialize data series
%--------------------------
leap=fix((sd(1)-366)/1460);            % leap years since 1900
ys=fix((sd(1)-leap-1)/365)+1900;       % year starting data series
sds=(ys-1900)*365+leap+1;              % serial day starting year
```

Appendix 6: Current analysis and current prediction programs

```
jd=sd-ones(n,1)*sds(1);                     % Julian day plus fractional day
jt=jd*24;                                    % Julian time in hours
st=sd*24;                                    % serial time in hours since 1900
sr=round(1/((jd(2)-jd(1))*24));              % sampling rate (no. readings per hour)
Ne=24*sera*sr;                               % series length expected for analysis
%-------------------------------
% find first whole Julian day
%-------------------------------
c=mod(jd(1),1)>(0.0416/sr);
jdf=fix(jd(1))+c; dte=fix(sd(1))+datenum('30-Dec-1899');
first_record=[datestr(dte),' ',num2str(jd(1))]
jds=jdf; jdlast=fix(jd(n)); jdsec=jdlast-round(Ne/(24*sr));
%---------------------
% start analysis loop
% select starting day
%---------------------
while jds>0
  vnum=0; SRate_Start1_Start2_lastday=[sr jdf jdsec jdlast]
  while vnum==0
    vnum=1;
    jds=input('Enter Julian Day to start analysis <0 to quit>: ');
    if jds==0 break; end
    jde=jds+Ne/(24*sr);
    if jds<jdf vnum=0; elseif jde>jdlast vnum=0; end
  end
  if jds==0 break; end
%--------------------
% find jds,jde indices
%--------------------
for i=1:n
  if jd(i)>=jds is=i; break; end
end
for i=1:n
  if jd(i)>=jde ie=i-1; break; end
end
N=ie-is+1;
Records_expected_found=[Ne N]             % number of records expected _ number found
t=st(is:ie)';                              % serial time vector selected for analysis
ajt=jt(is:ie)';                            % Julian time vector in hours
if spd==1

%-------------------------------
% convert S&D to UV form
%-------------------------------
   U=X(is:ie,3).*sin(X(is:ie,4)*pi/180);
   V=X(is:ie,3).*cos(X(is:ie,4)*pi/180);
 else
   U=X(is:ie,2); V=X(is:ie,3);
```

Appendix 6: Current analysis and current prediction programs

```
end
UV=[U V]; UVmean=mean(UV);  % UV current mean
%------------------
% UV scatterplot
%------------------
figure
plot(U,V,'r.',UVmean(1),UVmean(2),'k+')
title(file); xlabel(us); ylabel(vs);
axis equal
VV=axis; Rp=0.95*max(abs(VV));
  hold on
%---------------------------------------------------------------------
% eigenanalysis to determine principal axes of UV current field
%---------------------------------------------------------------------
UVd=UV-ones(N,1)*UVmean;            % UV deviate form
VCV=UVd'*UVd/(N-1);                  % variance-covariance matrix
[E1 E2 Q]=svd(VCV); TR=trace(E2);    % singular value decomposition
PAV=[E2(1,1)/TR E2(2,2)/TR];         % principal axis variance
Principal_Axis_Variance=PAV
TanTheta=E1(2,1)/E1(1,1);
MVDA=atan(TanTheta)*180/pi;  % maximum variance deviation angle (degCCW)
%----------------------------------------------------
% display principal axis headings (degTRUE)
%----------------------------------------------------
Principal_Axis_Headings=[90-MVDA 270-MVDA]
  mv=MVDA*pi/180;
  x(1)=Rp*cos(mv)+UVmean(1); y(1)=Rp*sin(mv)+UVmean(2);
  x(2)=Rp*cos(mv+pi)+UVmean(1); y(2)=Rp*sin(mv+pi)+UVmean(2);
  plot(x,y);
  text(VV(1)*0.92,VV(4)*0.9,sprintf('%5.1f',(270-MVDA)))
  text(VV(2)*0.79,VV(3)*0.9,sprintf('%5.1f',(90-MVDA)))
  plot(VV(1:2),[0 0],'k--',[0 0],VV(3:4),'k--')
  hold off
%-----------------------------
% Choose new major axis
%-----------------------------
  east=input('Change U-positive axis heading from 90 (east) to <0-360>: ','s');
  ra=(90.0-str2num(east))*pi/180;
  UVdro=UVd*[cos(ra) -sin(ra); sin(ra) cos(ra)];      % rotated UVd matrix
  UVmro=UVmean*[cos(ra) -sin(ra); sin(ra) cos(ra)];   % rotated UV means
  UVro=UVdro+ones(N,1)*UVmro;                % rotated UV matrix
  Uro=UVro(:,1); Vro=UVro(:,2);              % rotated U,V vectors
%-----------------------------------------------------
% set up sums of squares matrix for N points
% solve normal equations for least squares fit
%-----------------------------------------------------
  A=ones(N,1);
    for j=1:NC
```

Appendix 6: Current analysis and current prediction programs

```
      A=[A cos(w(j)*t)' sin(w(j)*t)'];
   end
  a=(A'*A)\(A'*UVdro);
%------------------------------------------
% determine UV amplitude and phase
% for NC harmonic constituents
%------------------------------------------
   for j=1:NC
      Ru(j)=sqrt(a(2*j,1)^2+a(2*j+1,1)^2);
      Rv(j)=sqrt(a(2*j,2)^2+a(2*j+1,2)^2);
      Zu(j)=atan2(a(2*j+1,1),a(2*j,1));
      Zv(j)=atan2(a(2*j+1,2),a(2*j,2));
   end
   Ruo=a(1,1); Rvo=a(1,2);
%------------------------
% predict UV currents
%------------------------
   Up=Ruo+Ru*cos(w'*t-Zu'*ones(1,N));     % predicted U component for series
   Vp=Rvo+Rv*cos(w'*t-Zv'*ones(1,N));     % predicted V component for series
   Upr=(Up'+ones(N,1)*UVmro(1))';         % predicted U component plus mean
   Vpr=(Vp'+ones(N,1)*UVmro(2))';         % predicted V component plus mean
   UVpr=[Upr' Vpr']; UVp=[Up' Vp'];
   Constituents
   Pudeg=mod(Zu*180/pi,360);
   Pvdeg=mod(Zv*180/pi,360);
%-----------------------------
% display LSQ fit statistics
%-----------------------------
   rms=sqrt((UVro-UVpr)'*(UVro-UVpr)/N);   % UV RMS error
   UV_rms_error=[rms(1,1) rms(2,2)]
   PVO=UVdro'*UVdro/(N-1);
   PVP=UVp'*UVp/(N-1);
   pv=PVP*100/PVO;                         % Percent of total variance accounted for
   UV_VarianceReduction=[pv(1,1) pv(2,2)]  % Reduction in variance per axis
%-----------------------------------------------
% compute orbital elements
% convert R to H (18.6-yr mean_current)
% convert Pdeg to nodal origin
%-----------------------------------------------
   jdb=fix((ajt(1)/24));      % Julian day beginning series
   Di=jdb+leap;               % First day + leapyear days since 1901
%-----------------------------------------------------
% mean longitude of the ascending lunar node
   Nm=259.16-19.3282*(ys-1900)-0.053*Di;
%-----------------------------------------------------
   Nmr=Nm*pi/180; fM2=1-0.037*cos(Nmr);
   fK1=1.006+0.115*cos(Nmr)-0.009*cos(2*Nmr);
   fO1=1.009+0.187*cos(Nmr)-0.015*cos(2*Nmr);
```

196 Appendix 6: Current analysis and current prediction programs

```
  Hu(1)=Ru(1)/fM2; Hu(2)=Ru(2); Hu(3)=Ru(3)/fM2;
  Hv(1)=Rv(1)/fM2; Hv(2)=Rv(2); Hv(3)=Rv(3)/fM2;
  Hu(4)=Ru(4)/fK1; Hu(5)=Ru(5)/fO1;
  Hv(4)=Rv(4)/fK1; Hv(5)=Rv(5)/fO1;
  Hu(6)=Ru(6)/(fM2^2); Hu(7)=Ru(7)/(fM2^3);
  Hv(6)=Rv(6)/(fM2^2); Hv(7)=Rv(7)/(fM2^3);
  Hu(8)=Ru(8); Hu(9)=Ru(9)/fM2;
  Hv(8)=Rv(8); Hv(9)=Rv(9)/fM2;
  uM2=-2.1*sin(Nmr); uK1=-8.9*sin(Nmr)+0.7*sin(2*Nmr);
  uO1=10.8*sin(Nmr)-1.3*sin(2*Nmr)+0.2*sin(3*Nmr);
  Pudeg(1)=Pudeg(1)+uM2; Pudeg(3)=Pudeg(3)+uM2;
  Pvdeg(1)=Pvdeg(1)+uM2; Pvdeg(3)=Pvdeg(3)+uM2;
  Pudeg(4)=Pudeg(4)+uK1; Pudeg(5)=Pudeg(5)+uO1;
  Pvdeg(4)=Pvdeg(4)+uK1; Pvdeg(5)=Pvdeg(5)+uO1;
  Pudeg(6)=Pudeg(6)+2*uM2; Pudeg(7)=Pudeg(7)+3*uM2;
  Pvdeg(6)=Pvdeg(6)+2*uM2; Pvdeg(7)=Pvdeg(7)+3*uM2;
  Pudeg(9)=Pudeg(9)+uM2; Pvdeg(9)=Pvdeg(9)+uM2;
%-----------------------------------
% display amplitude and phase
% display UV means
%-----------------------------------
  U_rotated_Amplitude_Phase=[Hu;Pudeg]      % U-rotated amplitude and phase(degs)
  V_rotated_Amplitude_Phase=[Hv;Pvdeg]      % V-rotated amplitude and phase(degs)
  UVmeans=UVmro                             % rotated UV means
%-----------------------------------
% store results in new spreadsheet
%-----------------------------------
  fout=fopen('hamels_current_out.xls','w');  % opening data outfile
  fprintf(fout,'%s',[file(1:4),' ',units]);  % writing header info
  fprintf(fout,'%s',['  Y',file(5:8)]);
  fprintf(fout,'%s\n',num2str(jds)); j=1;
  for i=1:NC
    tc=Constituents(j:j+3); j=j+3;           % write results to outfile
    fprintf(fout,'%s',tc);
    fprintf(fout,'%8.4f %8.4f\n',Hu(i),Pudeg(i));
  end
  uhd=str2num(east); uhd=round(uhd);
  fprintf(fout,'%s','Paxis');
  fprintf(fout,'%7.3f %7.2f\n',UVmro(1),uhd); % write Urmean, PAxis heading
  fprintf(fout,'%3.0f',uc);                   % write units: kts or mps
  fprintf(fout,'%s\n','  Umean Uheading');
  status=fclose(fout);
%--------------
% main plots
%--------------
  figure
  jdp=ajt(1:N)/24; Vm=UVmro(2)*ones(1,N); vmx=max(Vro); vhd=uhd-90;
  if vhd<0 vhd=vhd+360; end
```

```
    plot(jdp,Vro','r-',jdp,Vpr,'b-',jdp,Vro'-Vpr,'g-',jdp,Vro'*0,'k--',jdp,Vm,'k-')
    title(file)
    legend('Observed','Predicted','Residual')
    xlabel('Julian Day')
    ylabel(vns)
    text(jds+0.2,vmx-0.04,['Vp+ heading: ',num2str(vhd)])
    figure
    jdp=ajt(1:N)/24; Um=UVmro(1)*ones(1,N); umx=max(Uro);
    plot(jdp,Uro','r-',jdp,Upr,'b-',jdp,Uro'-Upr,'g-',jdp,Uro'*0,'k--',jdp,Um,'k-')
    title(file)
    legend('Observed','Predicted','Residual')
    xlabel('Julian Day')
    ylabel(uns)
    text(jds+0.2,umx+0.04,['Up+ heading:',num2str(uhd)])
%-------------------
% three-day plots
%-------------------
    p=1; jmax=jde-4;
    while p>0
      vnum=0;
      while vnum==0
        p=input('enter Julian day for three-day plot <0 to exit>: ');
        vnum=1;
        if p==0 break; end
        if p<jds vnum=0;
        elseif p>jmax vnum=0; end
      end
      if p==0 break; end
      j=(p-jds)*24*sr+1; k=72*sr-1; jt3=ajt(j:j+k)/24;
      Uo3=Uro(j:j+k); Up3=Upr(j:j+k); Vo3=Vro(j:j+k); Vp3=Vpr(j:j+k);
      figure
      plot(jt3,Vo3','r-',jt3,Vp3,'b-',jt3,Vo3'-Vp3,'g-',jt3,Vo3'*0,'k--')
      title(file)
      xlabel('Julian Day')
      ylabel(vns)
      figure
      plot(jt3,Uo3','r-',jt3,Up3,'b-',jt3,Uo3'-Up3,'g-',jt3,Uo3'*0,'k--')
      title(file)
      xlabel('Julian Day')
      ylabel(uns)
    end
  end
harmonic_constants_file=['hamels_current_out.xls']
```

ASTRO_CURRENT.m

```
clear
echo on
```

Appendix 6: Current analysis and current prediction programs

```
% ASTRO_CURRENT            - Astronomical Water Current Prediction
% generates currents outfile - 'astro_current_out.xls'
% Time origin for prediction - Midnight ending December 30,1899 LST
% Tidal harmonic constituents - NC constituents from HAMELS_CURRENT
% Numeric Input (X matrix)  - NC rows of amplitude(m) and phase(deg)
% Numeric Input (X matrix)  - one row, current mean and principal direction
% Version 2.1 for EXCEL     - written by J.D. Boon (VIMS) 12/5/2002
echo off
NC=9;                       % Number of tidal harmonic constituents
m2=28.9841042*pi/180;       % M2 frequency (deg/mean solar hour)
s2=30.0000000*pi/180;       % S2 frequency (deg/mean solar hour)
n2=28.4397295*pi/180;       % N2 frequency (deg/mean solar hour)
k1=15.0410686*pi/180;       % K1 frequency (deg/mean solar hour)
o1=13.9430356*pi/180;       % O1 frequency (deg/mean solar hour)
m4=2*m2;                    % M4 frequency (deg/mean solar hour)
m6=3*m2;                    % M6 frequency (deg/mean solar hour)
s4=2*s2;                    % S4 frequency (deg/mean solar hour)
ms4=m2+s2;                  % MS4 frequency(deg/mean solar hour)
w=[m2 s2 n2 k1 o1 m4 m6 s4 ms4]; % frequency vector
%----------------------------------
% Enter harmonic constants file
%----------------------------------
file=input('Enter harmonic constants filename <e.g., ChesBayEntrance19m.xls>: ','s');
[X,A]=xlsread(file); [n,m]=size(X);
H=X(1:n-2,2); P=X(1:n-2,3); R=H; Pdeg=P;
Um=X(n-1,2); PD=X(n-1,3); eu=X(n,1); cf=1;
 year=input('   Enter year of predictions <yyyy>: ','s');
   nu=input('    current spd in kts(1) or mps(2) : ');
tpred=input('    plot one day(1),month(2),year(3): ');
if nu<2
    units='U(knots)';
    if eu>1 cf=1.9438445; end
else units='U(mps)';
    if eu<2 cf=1/1.9438445; end
end
bd=['01/01/',year]; bdv=datevec(bd); spc=' '; ys=str2num(year);
%---------------------------------------------
% compute orbital elements
% convert R to H (18.6-yr mean)
% convert Pdeg to nodal origin
%------------------------------------------------------------------------
 Di=fix((ys-1901)/4)+183; % leapyear days since 1901 +183
%------------------------------------------------------------------------
% mean longitude of the ascending lunar node
 Nm=259.16-19.3282*(ys-1900)-0.053*Di;
%------------------------------------------
 Nmr=Nm*pi/180; fM2=1-0.037*cos(Nmr);
 fK1=1.006+0.115*cos(Nmr)-0.009*cos(2*Nmr);
```

Appendix 6: Current analysis and current prediction programs

```
fO1=1.009+0.187*cos(Nmr)-0.015*cos(2*Nmr);
R(1)=H(1)*fM2; R(3)=H(3)*fM2;
R(4)=H(4)*fK1; R(5)=H(5)*fO1;
R(6)=H(6)*(fM2^2); R(7)=H(7)*(fM2^3); R(9)=H(9)*fM2;
uM2=-2.1*sin(Nmr); uK1=-8.9*sin(Nmr)+0.7*sin(2*Nmr);
uO1=10.8*sin(Nmr)-1.3*sin(2*Nmr)+0.2*sin(3*Nmr);
Pdeg(1)=Pdeg(1)+uM2; Pdeg(3)=Pdeg(3)+uM2;
Pdeg(4)=Pdeg(4)+uK1; Pdeg(5)=Pdeg(5)+uO1;
Pdeg(6)=Pdeg(6)+2*uM2; Pdeg(7)=Pdeg(7)+3*uM2;
Pdeg(9)=Pdeg(9)+uM2; P=Pdeg*pi/180;
if tpred==1
  ss(1:19)=spc; sdat=input([ss,'month/day <mm/dd>: '],'s');
  sdv=datevec(sdat); sdv(1)=str2num(year);
  sd=datenum(sdv)-datenum('30-Dec-1899'); ed=sd+1;
  ti=0.5; t=[sd*24:ti:ed*24-ti]; N=length(t); tp=t/24;
  Up=Um*ones(1,N)+R'*cos(w'*t-P*ones(1,N)); Up=Up*cf;
  stamp=[file(1:4),' ',year,' ',sdat]; umx=max(Up);
  figure
  plot(tp,Up,'b-',tp,Up*0,'k--')
  datetick('x',16)
  title(stamp)
  xlabel('Local Standard Time (hours)')
  ylabel(units)
  text(tp(2),umx+0.04,'U+ heading:')
  text(tp(10),umx+0.04,num2str(PD))
  grid on
elseif tpred==2
  ss(1:23)=spc; ms=input([ss,'month <1..12>: ']); me=ms+1;
  bdv(2)=ms; sd=datenum(bdv)-datenum('30-Dec-1899');
  bdv(2)=me; ed=datenum(bdv)-datenum('30-Dec-1899');
  ti=1; t=[sd*24:ti:ed*24-ti]; N=length(t); tp=t/24;
  Up=Um*ones(1,N)+R'*cos(w'*t-P*ones(1,N)); Up=Up*cf;
  stamp=[file(1:4),' ',year]; umx=max(Up);
  figure
  plot(tp,Up,'b-',tp,Up*0,'k--')
  datetick('x',6)
  title(stamp)
  xlabel('Local Standard Time (days)')
  ylabel(units)
  text(tp(10),umx+0.1,'U+ heading:')
  text(tp(140),umx+0.1,num2str(PD))
elseif tpred==3
  ms=1; me=ms+12;
  bdv(2)=ms; sd=datenum(bdv)-datenum('30-Dec-1899');
  bdv(2)=me; ed=datenum(bdv)-datenum('30-Dec-1899');
  ti=1; t=[sd*24:ti:ed*24-ti]; N=length(t); tp=t/24;
  Up=Um*ones(1,N)+R'*cos(w'*t-P*ones(1,N)); Up=Up*cf; umx=max(Up);
  stamp=[file(1:4),' ',year]; tx=tp-(sd-1)*ones(1,N);
```

```
    figure
    plot(tx,Up,'b-',tx,Up*0,'k--')
    V=axis; V(1)=1; V(2)=366;
    axis(V)
    if mod(str2num(year),4)==0
        set(gca,'xtick',[1 30 60 90 120 150 180 210 240 270 300 330 366])
    else
        set(gca,'xtick',[1 30 60 90 120 150 180 210 240 270 300 330 365])
    end
    title(stamp)
    xlabel('Julian Days')
    ylabel(units)
    text(tx(80),umx+0.14,'U+ heading:')
    text(tx(1600),umx+0.14,num2str(PD))
end
sav=input('Save this data? <y/n>: ','s');
if sav=='y'
    saved_file=['astro_current_out.xls']
    fout=fopen('astro_current_out.xls','w');      % opening data outfile
    fprintf(fout,'%s',file(1:4));                 % writing header info
    fprintf(fout,'%s',[year,'    ',units]);
    fprintf(fout,'%s\n',[' ',num2str(PD)]);
    data=[tp' Up'];
    fprintf(fout,'%10.5f %9.3f\n',data');          % writing data to file
    status=fclose(fout);
end
```

References

CHAPTER 1

Books

Pickard, G.L. and Emery, W.J., 1982, Descriptive physical oceanography: an introduction, 4th Ed., Pergamon Press.

Redfield, A.C., 1980, *Introduction to tides: the tides of the waters of New England and New York,* Marine Science International, Woods Hole, MA.

Wilstach, P., 1929, *Tidewater Virginia*, Tudor Publishing Company, New York.

Papers and Special Publications

Kuo, A. Y. and Neilson, B.J., 1987, 'Hypoxia and salinity in Virginia estuaries', *Estuaries*, **10**, 277-283.

Simpson, J.H., 1998, 'Tidal processes in shelf seas', *The Sea*, **10**, Ch. 5, 113-150.

CHAPTER 2

Books

Cartwright, D.E., 1999, *Tides: a scientific history*, Cambridge University Press.

Defant, A., 1958, *Ebb and flow: the tides of earth, air, and water*, The University of Michigan Press, Ann Arbor, MI.

Oceanography Course Team, 1989, *Waves, tides and shallow-water processes*, The Open University, Pergamon Press.

Government publications

Doodson, A.T. and Warburg, H.D., 1941, *Admiralty manual of tides*, Publication NP120 reprinted 1980, Admiralty Charts and Publications, London.

CHAPTER 3

Books

Cartwright, D.E. (see Chapter 2 reference list).

Dean, R.G. and Dalrymple, R.A., 1984, *Water wave mechanics for engineers and scientists*, Prentice-Hall, Englewood Cliffs, NJ.

Goda, Y., 1985, *Random seas and design of maritime structures*, University of Tokyo Press.

Mellor, G.L., 1996, *Introduction to physical oceanography*, American Institute of Physics Press, Woodbury, NY.

Historical Reference

Harris, R.A., 1904, *Cotidal lines of the world*, Manual of Tides, Appendix No. 5, US Government Printing Office, Washington.

CHAPTER 4

Books

Defant, A. (see Chapter 2 reference list).

Doodson, A.T. (see Chapter 2 reference list).

Pugh, D.T., 1987, *Tides, surges and mean sea-level*, John Wiley & Sons.

CHAPTER 5

Books

Emery, K.O. and Aubrey, D.G., 1991, *Sea levels, land levels and tide gauges*, Springer-Verlag, New York.

Pugh, D.T., 2004, *Changing sea levels: effects of tides, weather and climate*, Cambridge University Press.

Government Publications

Doodson, A.T. (see Chapter 2 reference list).

Hatfield, H.R., 1969, *Admiralty manual of hydrographic surveying: tides and tidal streams,* Publication NP134b(2) reprinted 1983, Admiralty Charts and Publications, London.

Hicks, S.D., 1984, *Tide and current glossary*, US Department of Commerce, National Oceanic and Atmospheric Administration, National Ocean Service.

Lyles, S.D., Hickman, L.E. and Debaugh, H.A., 1988, *Sea level variations for the United States 1855-1986*, Office of Oceanography and Marine Assessment, National Oceanic and Atmospheric Administration, National Ocean Service.

CHAPTER 6

Books

Oceanography Course Team (see Chapter 2 reference list).

Pickard, G.L. and Emery, W.J. (see Chapter 1 reference list).

Redfield, A.C., 1958, 'The influence of the continental shelf on the tides of the Atlantic Coast of the United States', *Journal of Marine Research*, **17**, 432-448.

Government Publications

Browne, D.R. and Fisher, C.W., 1988, *Tide and tidal currents in the Chesapeake Bay*, Technical Report NOS OMA3, Office of Oceanography and Marine Assessment, National Oceanic and Atmospheric Administration, National Ocean Service.

CHAPTER 7

Books

Bloomfield, P., 1976., *Fourier analysis of time series: an introduction*, John Wiley & Sons.

Papers
Cheng, R.T., McKinnie, D., English, C. and Smith, R.E., 1998, 'An overview of San Francisco Bay PORTS', *Ocean Community Conference 98*, Marine Technology Society, 1054-1060.

Wang, D.P., 1979, 'Subtidal sea level variations in the Chesapeake Bay and relations to atmospheric forcing', *Journal of Physical Oceanography*, 413-421.

CHAPTER 8

Papers

Cheng, R.T. and Smith, R.E., 1998, 'A nowcast model for tides and tidal currents in San Francisco Bay, California', *Ocean Community Conference 98*, Marine Technology Society, 1054-1060.

Government Publications

Doodson, A.T. (see Chapter 2 reference list).

Schureman, P., *Manual of harmonic analysis and prediction of tides*. Special Publication No. 98 reprinted 1958, US Department of Commerce, Coast and Geodetic Survey.

CHAPTER 9

Books

Cushman-Roisin, B., Miroslav G., Poulain, P-M., and Artegiani, A. (ed.), 2001, *Physical Oceanography of the Adriatic Sea: Past, Present and Future*, Ch. 5, Sec. 5.2, Ch. 6, Sec. 6.3.3, Kluwer Academic Publishers, Dordrecht.

Pugh, D.T. (see Chapter 4 reference list).

Papers and Special Publications

Jelesnianski, C.P., 1993, 'The Habitation Layer: Storm Surges', *The Global Guide to Tropical Cyclone Forecasting WMO/TD-560* (Ch. 4), World Meteorological Organization, Geneva.

Groen, P. and Groves, G.W., 1962, 'Surges', *The Sea*, **1**, Ch.17, 611-646.

Redfield, A.C. (see Chapter 6 references list).

CHAPTER 10

Bloomfield, P. (see Chapter 7 reference list).

Glossary

The following definitions are excerpted from the *Tide and Current Glossary* (1999 edition) published in the United States by the National Ocean Service, National Oceanic and Atmospheric Administration., US Department of Commerce.

acoustic Doppler current profiler (ADCP) – current measuring instrument employing the transmission of high frequency acoustic signals in the water. The current is determined by a Doppler shift in the backscatter echo from plankton, suspended sediment, and bubbles, all assumed to be moving with the mean speed of the water. Time gating circuitry is employed which uses differences in acoustic travel time to divide the water column into range intervals, called bins. The bin determinations allow development of a profile of current speed and direction over most of the water column. The ADCP can be deployed from a moving vessel, tow, buoy, or bottom platform.

amphidromic point - A point of zero amplitude of the observed or a constituent tide.

amphidromic region - An area surrounding an amphidromic point from which the radiating cotidal lines progress through all hours of the tidal cycle.

amplitude (H) - One-half the range of a constituent tide. By analogy, it may be applied also to the maximum speed of a constituent current.

analysis, harmonic—See harmonic analysis.

aphelion - The point in the orbit of the Earth (or other planet, etc.) farthest from the Sun.

apogean tides or tidal currents - Tides of decreased range or currents of decreased speed occurring monthly as the result of the Moon being in apogee.

apogee - The point in the orbit of the Moon or a man-made satellite farthest from the Earth. The point in the orbit of a satellite farthest from its companion body.

apparent secular trend - The non-periodic tendency of sea level to rise, fall, or remain stationary with time. Technically, it is frequently defined as the slope of a least squares line of regression through a relatively long series of yearly mean sea-level values. The word 'apparent' is used since it is often not possible to know whether a trend is truly non-periodic or merely a segment of a very long oscillation (relative to the length of the series).

barycenter - The common center of mass of the Sun-Earth System or the Moon-Earth System. The distance from the center of the Sun to the Sun-Earth barycenter is about 280 miles. The distance from the center of the Earth to the Moon-Earth barycenter is about 2,895 miles.

bench mark (BM) - A fixed physical object or mark used as reference for a horizontal or vertical datum. A tidal bench mark is one near a tide station to which the tide staff and tidal datums are referred. A primary bench mark is the principal mark of a group of tidal bench marks to which the tide staff and tidal datums are referred.

celestial sphere - An imaginary sphere of infinite radius concentric with the Earth on which all celestial bodies except the Earth are imagined to be projected.

chart datum - The datum to which soundings on a chart are referred. It is usually taken to correspond to a low-water elevation (expressed as a depression below mean sea level).

cocurrent line - A line on a map or chart passing through places having the same current hour.

comparison of simultaneous observations - A reduction process in which a short series of tide or tidal current observations at any place is compared with simultaneous observations at a control station where tidal or tidal current constants have previously been determined from a long series of observations. The observations are typically high and low tides and monthly means. For tides, it is usually used to adjust constants from a subordinate station to the equivalent value that would be obtained from a 19-year series.

compound tide - A harmonic tidal (or tidal current) constituent with a speed equal to the sum or difference of the speeds of two or more elementary constituents. The presence of compound tides is usually attributed to shallow water conditions.

corange line - A line passing through places of equal tidal range.

Coriolis force - A fictional force in the hydrodynamic equations of motion that takes into account the effect of the Earth's rotation on moving objects (including air and water) when viewed with reference to a coordinate system attached to the rotating Earth. The horizontal component is directed 90° to the right (when looking in the direction of motion) in the Northern Hemisphere and 90° to the left in the Southern. The horizontal component is zero at the Equator; also, when the object is at rest relative to the Earth.

cotidal hour - The average interval between the Moon's transit over the meridian of Greenwich and the time of the following high water at any place. This interval may be expressed either in solar or lunar time. When expressed in solar time, it is the same as the Greenwich high water interval.

cotidal line - A line on a chart or map passing through places having the same tidal hour.

current - Generally, a horizontal movement of water. Currents may be classified as tidal and non-tidal. Tidal currents are caused by gravitational interactions between the Sun, Moon, and Earth and are part of the same general movement of the sea that is manifested in the vertical rise and fall, called tide. Tidal currents are periodic with a net velocity of zero over the particular tidal cycle.

current ellipse - A graphic representation of a rotary current in which the velocity of the current at different hours of the tidal cycle is represented by radius vectors and vectoral angles. A line joining the extremities of the radius vectors will form a curve roughly approximating an ellipse. The cycle is completed in one-half tidal day or in a whole tidal day, according to whether the tidal current is of the semidiurnal or the diurnal type.

current hour - The mean interval between the transit of the Moon over the meridian of Greenwich and the time of strength of flood, modified by the times of slack water (or minimum current) and strength of ebb.

datum (vertical) - For marine applications, a base elevation used as a reference from which to reckon heights or depths. It is called a tidal datum when defined in terms of a certain phase of the tide. Tidal datums are local datums and should not be extended into areas which have differing hydrographic characteristics without substantiating measurements.

declination (cycle) - Angular distance north or south of the celestial equator, taken as positive when north of the equator and negative when south. The Sun passes through its declinational cycle once a year, reaching its maximum north declination of approximately 23-½° about June 21 and its maximum south declination of approximately 23-½° about December 21. The Moon has an average declinational cycle of 27-⅓ days which is called a tropical month.

diurnal - Having a period or cycle of approximately one tidal day. Thus, the tide is said to be diurnal when only one high water and one low water occur during a tidal day, and the tidal current is said to be diurnal when there is a single flood and a single ebb period of a reversing current in the tidal day. A rotary current is diurnal if it changes its direction through all points of the compass once each tidal day.

diurnal inequality - The difference in height of the two high waters or of the two low waters of each tidal day; also, the difference in speed between the two flood tidal currents or the two ebb currents of each tidal day.

ebb current (ebb) - The movement of a tidal current away from shore or down a tidal river or estuary.

ecliptic - The intersection of the plane of the Earth's orbit with the celestial sphere.

epoch - (1) Also known as phase lag. Angular retardation of the maximum of a constituent of the observed tide (or tidal current) behind the corresponding maximum of the same constituent of the theoretical equilibrium tide. (2) As used in tidal datum determination, it is a 19-year cycle over which tidal height observations are meaned in order to establish the various datums. As there are periodic and apparent secular trends in sea level, a specific 19-year cycle (the National Tidal Datum Epoch) is selected so that all tidal datum determinations throughout the United States, its territories, Commonwealth of Puerto Rico, and Trust Territory of the Pacific Islands, will have a common reference.

equatorial tides - Tides occurring semimonthly as a result of the Moon being over the Equator.

At these times the tendency of the Moon to produce a diurnal inequality in the tide is at a minimum.

equilibrium argument - The theoretical phase of a constituent of the equilibrium tide. It is usually represented by the expression $(V + u)$, in which V is a uniformly changing angular quantity involving multiples of the hour angle of the mean Sun, the mean longitudes of the Moon and Sun, and the mean longitude of lunar or solar perigee; and u is a slowly changing angle depending upon the longitude of the Moon's node. When pertaining to an initial instant of time, such as the beginning of a series of observations, it is expressed by $(Vo+ u)$.

equilibrium theory - A model under which it is assumed that the waters covering the face of the Earth instantly respond to the tide-producing forces of the Moon and Sun to form a surface of equilibrium under the action of these forces. The model disregards friction, inertia, and the irregular distribution of the land masses of the earth. The theoretical tide formed under these conditions is known as the equilibrium tide.

equinoctial tides - Tides occurring near the times of the equinoxes.

equinoxes - The two points in the celestial sphere where the celestial equator intersects the ecliptic; also, the times when the Sun crosses the equator at these points. The vernal equinox is the point where the Sun crosses the Equator from south to north and it occurs about March 21. Celestial longitude is reckoned eastward from the vernal equinox. The autumnal equinox is the point where the Sun crosses the Equator from north to south and it occurs about September 23.

estuary - An embayment of the coast in which fresh river water entering at its head mixes with the relatively saline ocean water. When tidal action is the dominant mixing agent it is usually termed a tidal estuary. Also, the lower reaches and mouth of a river emptying directly into the sea where tidal mixing takes place. The latter is sometimes called a river estuary.

flood current (flood) - The movement of a tidal current toward the shore or up a tidal river or estuary.

flow - The British equivalent of the United States total current. Flow is the combination of tidal stream and current.

flushing time - The time required to remove or reduce (to a permissible concentration) any dissolved or suspended contaminant in an estuary or harbor.

forced wave—A wave generated and maintained by a continuous force.

free wave - A wave that continues to exist after the generating force has ceased to act.

geopotential (equipotential) surface - A surface that is everywhere normal to the acceleration of gravity.

Greenwich interval - An interval referred to the transit of the Moon over the meridian of Greenwich, as distinguished from the local interval which is referred to the Moon's transit over the local meridian.

harmonic analysis - The mathematical process by which the observed tide or tidal current at any place is separated into basic harmonic constituents.

harmonic constants - The amplitudes and epochs of the harmonic constituents of the tide or tidal current at any place.

head of tide - The inland or upstream limit of water affected by the tide.

high water (HW) - The maximum height reached by a rising tide.

Indian spring low water - A datum originated by Professor G. H. Darwin when investigating the tides of India. It is an elevation depressed below mean sea level by an amount equal to the sum of the amplitudes of the harmonic constituents M2, S2, K1, and O1.

intertidal zone - (technical definition) The zone between the mean higher high water and mean lower low water lines.

inverse barometer effect - The inverse response of sea level to changes in atmospheric pressure. A static reduction of 1.005 mb in atmospheric pressure will cause a stationary rise of 1 cm in sea level.

Julian date - Technique for the identification of successive days of the year when monthly notation is not desired. This is especially applicable in computer data processing and acquisition where indexing is necessary.

knot - A speed unit of 1 international nautical mile (1,852.0 meters or 6,076.115,49 international feet) per hour.

leap year - A calendar year containing 366 days. According to the present Gregorian calendar, all years with the date-number divisible by 4 are leap years, except century years. The latter are leap years when the date-number is divisible by 400.

long period constituent - A tidal or tidal current constituent with a period that is independent of the rotation of the Earth but which depends upon the orbital movement of the Moon or the Earth. The principal lunar long period constituents have periods approximating a month and half a month, and the principal solar long period constituents have periods approximating a year and half a year.

low water (LW) - The minimum height reached by a falling tide.

lowest astronomical tide (LAT) - As defined by the International Hydrographic Organization, the lowest tide level that can be predicted to occur under average meteorological conditions and under any combination of astronomical conditions.

lunar day - The time of the rotation of the Earth with respect to the Moon, or the interval between two successive upper transits of the Moon over the meridian of a place. The mean lunar day is approximately 24.84 solar hours in length, or 1.035 times as great as the mean solar day.

lunar nodes - The points where the plane of the Moon's orbit intersects the ecliptic. The point where the Moon crosses in going from south to north is called the ascending node and the point where the crossing is from north to south is called the descending node.

lunitidal interval - The interval between the Moon's transit (upper or lower) over the local or Greenwich meridian and the following high or low water. The average of all high water intervals for all phases of the Moon is known as mean high water lunitidal interval and is abbreviated to high water interval (HWI). Similarly, mean low water lunitidal interval is abbreviated to low water interval (LWI).

mean high water (MHW) - A tidal datum. The average of all the high water heights of each tidal day observed over the National Tidal Datum Epoch. For stations with shorter series, comparison of simultaneous observations with a control tide station is made in order to derive the equivalent datum of the National Tidal Datum Epoch.

mean higher high water (MHHW) - A tidal datum. The average of the higher high water height of each tidal day observed over the National Tidal Datum Epoch. For stations with shorter series, comparison of simultaneous observations with a control tide station is made in order to derive the equivalent datum of the National Tidal Datum Epoch.

mean low water (MLW) - A tidal datum. The average of all the low water heights observed over the National Tidal Datum Epoch. For stations with shorter series, comparison of simultaneous observations with a control tide station is made in order to derive the equivalent datum of the National Tidal Datum Epoch.

mean lower low water (MLLW) - A tidal datum. The average of the lower low water height of each tidal day observed over the National Tidal Datum Epoch. For stations with shorter series, comparison of simultaneous observations with a control tide station is made in order to derive the equivalent datum of the National Tidal Datum Epoch.

mean range of tide (Mn) - The difference in height between mean high water and mean low water.

mean sea level (MSL) - A tidal datum. The arithmetic mean of hourly heights observed over the National Tidal Datum Epoch. Shorter series are specified in the name; e.g., monthly mean sea level and yearly mean sea level.

mean tide level (MTL) - A tidal datum. The arithmetic mean of mean high water and mean low water. Same as half-tide level.

meteorological tides - Tidal constituents having their origin in the daily or seasonal variations in weather conditions which may occur with some degree of periodicity. The principal meteorological constituents recognized in the tides are Sa, Ssa, and Sl. See storm surge.

National Tidal Datum Epoch - The specific 19-year period adopted by the National Ocean Service as the official time segment over which tide observations are taken and reduced to obtain mean values (e.g., mean lower low water, etc.) for tidal datums. It is necessary for standardization because of periodic and apparent secular trends in sea level. The present National Tidal Datum Epoch is 1983 through 2001. It is reviewed annually for possible revision and must be actively considered for revision every 25 years.

node cycle - Period of approximately 18.61 Julian years required for the regression of the Moon's nodes to complete a circuit of 360° of longitude. It is accompanied by a corresponding cycle of changing inclination of the Moon's orbit relative to the plane of the Earth's Equator, with resulting inequalities in the rise and fall of the tide and speed of the tidal current.

node factor (f) - A factor depending upon the longitude of the Moon's node which, when applied to the mean coefficient of a tidal constituent, will adapt the same to a particular year for which predictions are to be made.

overtide - A harmonic tidal (or tidal current) constituent with a speed that is an exact multiple of the speed of one of the fundamental constituents derived from the development of the tide-producing force. The presence of overtides is usually attributed to shallow water conditions.

perigean tides or tidal currents - Tides of increased range or tidal currents of increased speed occurring monthly as the result of the Moon being in perigee.

perigee - The point in the orbit of the Moon or a man-made satellite nearest to the Earth. The point in the orbit of a satellite nearest to its companion body.

perihelion - The point in the orbit of the Earth (or other planet, etc.) nearest to the Sun.

period - Interval required for the completion of a recurring event, such as the revolution of a celestial body or the time between two consecutive like phases of the tide or tidal current. A period may be expressed in angular measure as 360°. The word also is used to express any specified duration of time.

prime meridian - The meridian of longitude which passes through the original site of the Royal Observatory in Greenwich, England and used as the origin of longitude. Also known as the Greenwich Meridian.

progressive wave - A wave that advances in distance along the sea surface or at some intermediate depth. Although the wave form itself travels significant distances, the water particles that make up the wave merely describe circular (in relatively deep water) or elliptical (in relatively shallow water) orbits.

range of tide - The difference in height between consecutive high and low waters. The mean range is the difference in height between mean high water and mean low water.

red tide (water) - The term applied to toxic algal blooms caused by several genera of dinoflagellates (*Gymnodinium* and *Gonyaulax*) which turn the sea red and are frequently associated with a deterioration in water quality. The color occurs as a result of the reaction of a red pigment, peridinin, to light during photosynthesis. These toxic algal blooms pose a serious threat to marine life and are potentially harmful to humans. The term has no connection with astronomic tides. However, its association with the word 'tide' is from observations of its movement with tidal currents in estuarine waters.

residual current - The observed current minus the astronomical tidal current.

reversing current - A tidal current which flows alternately in approximately opposite directions with a slack water at each reversal of direction. Currents of this type usually occur in rivers and straits where the direction of flow is more or less restricted to certain channels.

rotary current - A tidal current that flows continually with the direction of flow changing through all points of the compass during the tidal period. Rotary currents are usually found offshore where the direction of flow is not restricted by any barriers. The tendency for the rotation in direction has its origin in the Coriolis force and, unless modified by local conditions, the change is clockwise in the Northern Hemisphere, counterclockwise in the Southern.

seiche - A stationary wave usually caused by strong winds and/or changes in barometric pressure. It is found in lakes, semi-enclosed bodies of water, and in areas of the open ocean.

slack water (slack) - The state of a tidal current when its speed is near zero, especially the moment when a reversing current changes direction and its speed is zero.

semidiurnal - Having a period or cycle of approximately one-half of a tidal day. The predominant type of tide throughout the world is semidiurnal, with two high waters and two low waters each tidal day. The tidal current is said to be semidiurnal when there are two flood and two ebb periods each day.

shallow water constituent - A short-period harmonic term introduced into the formula of tidal (or tidal current) constituents to account for the change in the form of a tide wave resulting from shallow water conditions. Shallow water constituents include the overtides and compound tides.

210 Glossary

standard time - A kind of time based upon the transit of the Sun over a certain specified meridian, called the time meridian, and adopted for use over a considerable area. With a few exceptions, standard time is based upon some meridian which differs by a multiple of 15° from the meridian of Greenwich.

stilling well - A vertical pipe with a relatively small opening (intake) in the bottom. It is used in a gauge installation to dampen short period surface waves while freely admitting the tide, other long period waves, and sea level variations; which can then be measured by a water level gauge sensor inside.

storm surge - The local change in the elevation of the ocean along a shore due to a storm. The storm surge is measured by subtracting the astronomic tidal elevation from the total elevation. It typically has a duration of a few hours. Since wind generated waves ride on top of the storm surge (and are not included in the definition), the total instantaneous elevation may greatly exceed the predicted storm surge plus astronomic tide. It is potentially catastrophic, especially on low lying coasts with gently sloping offshore topography. See storm tide.

storm tide - As used by the National Weather Service, NOAA, the sum of the storm surge and astronomic tide. See storm surge.

synodic month—The average period of the revolution of the Moon around the Earth with respect to the Sun, or the average interval between corresponding phases of the Moon. The synodic month is approximately 29.530,588 days in length.

tidal bore - A tidal wave that propagates up a relatively shallow and sloping estuary or river with a steep wave front. The leading edge presents an abrupt rise in level, frequently with continuous breaking and often immediately followed by several large undulations. An uncommon phenomenon, the tidal bore is usually associated with very large ranges in tide as well as wedge shaped and rapidly shoaling entrances.

tidal characteristics - Principally, those features relating to the time, range, and type of tide.

tidal current - A horizontal movement of the water caused by gravitational interactions between the Sun, Moon, and Earth. The horizontal component of the particulate motion of a tidal wave. Part of the same general movement of the sea that is manifested in the vertical rise and fall called tide. The United States equivalent of the British tidal stream.

tide staff - A water level gauge consisting of a vertical graduated staff from which the height of the water level can be read directly. It is called a fixed staff when secured in place so that it cannot be easily removed.

tide (water level) station—The geographic location at which tidal observations are conducted. Also, the facilities used to make tidal observations. These may include a tide house, tide (water level) gauge, tide staff, and tidal bench marks.

Tide Tables - Tables which give daily predictions of the times and heights of high and low waters. These predictions are usually supplemented by tidal differences and constants through which predictions can be obtained for numerous other locations.

tidewater - Water activated by the tide generating forces and/or water affected by the resulting tide, especially in coastal and estuarine areas. Also, a general term often applied to the land and water of estuarine areas formed by postglacial drowning of coastal plain rivers.

tide (astronomical) - The periodic rise and fall of a body of water resulting from gravitational interactions between Sun, Moon, and Earth. The vertical component of the particulate motion of a tidal wave. Although the accompanying horizontal movement of the water is part of the same phenomenon, it is preferable to designate this motion as tidal current.

total current - The combination of the tidal and non-tidal current. The United States equivalent of the British flow. See current.

tractive force - The horizontal component of a tide-producing force vector (directed parallel with level surfaces at that geographic location).

tropic currents - Tidal currents occurring semimonthly when the effect of the Moon's maximum declination is greatest. At these times the tendency of the Moon to produce a diurnal inequality in the current is at a maximum.

tsunami - A shallow water progressive wave, potentially catastrophic, caused by an underwater earthquake or volcano.

universal time (UT) – Same as Greenwich mean time (GMT).

Index

Acoustic Doppler Meter (ADM),	109
Adriatic Sea,	148
August 1933 hurricane,	140
amphidromic system,	42, 103
apparent sun,	19
apogean, perigean tides,	16, 50
ascending lunar node,	20
Ash Wednesday storm,	137
astronomical tide,	93, 134
British Oceanographic Data Centre,	96
buildup factors,	128
Caribbean Sea tides,	43
celestial sphere, equator,	19
centrifugal force,	10
Chesapeake Bay Entrance currents,	112
Chesapeake Bay Entrance tides,	97
compound harmonic,	120
compound tides,	59, 102
CO-OPS (US NOAA/NOS),	96
co-range, co-tidal charts,	74
Coriolis effect,	39
critical resonance length,	81
current curve,	111
current ellipse,	87
deepwater wave,	28
differential forces,	11
diurnal inequality,	15, 55
ecliptic,	19
eigenvector analysis,	87
Electromagnetic Current Meter (ECM),	109
Elizabeth River estuary,	80
equilibrium tide,	13, 124
estuarine circulation,	5, 89
fall line,	1
Fast Fourier Transform (FFT),	100
forced, free waves,	7, 27
flood, ebb,	1
geoid,	61
geodetic datum,	61
global warming,	66
gravity,	8
Greenwich mean time,	53
harmonic constants,	125
harmonic expansion,	120
harmonic method of tidal analysis,	46, 94
harmonic tidal constituents,	46, 93
hurricane *Camille*,	136
hurricane *Isabel*,	139
hypoxia, anoxia,	7
intertidal zone,	3
Julian day (date),	96, 126
Kelvin machine,	57
Kelvin (rotary) wave,	33, 45
law of universal gravitation,	9
Liverpool Bay (UK) tides,	116
Liverpool, Mersey R (UK) tides,	100
local standard time,	53
long waves,	32
lowest astronomical tide,	54
low-pass digital filter,	94
lunar declinational cycle,	15, 18
lunar elliptic tide,	50
lunar semidiurnal tide,	47
lunar-solar semidiurnal tide,	95
lunar tides,	13
lunitidal interval,	14
MATLAB GUI *SIMPLY TIDES*,	126
MATLAB GUI *SIMPLY CURRENYS*,	131
mean high water,	54, 60, 68
mean low water,	54, 60, 68
mean lower low water,	54, 68
mean higher high water,	68, 143
mean sea level,	60, 68
mean tide level,	68
mean tide range,	68
mean water level,	34
method of simultaneous comparisons,	67
Meton's cycle,	63
microtidal, macrotidal,	2

Index

natural period,	37
nodal factors,	124
non-tidal currents,	88
Northern Adriatic storm tides,	148
Northwest Persian Gulf tides,	103
November 1966 storm (Venice, Italy),	149
nowcasts (data assimilation),	133
ocean tides,	33
overtides,	56, 59, 84, 101, 102
perihelion, aphelion,	20
Physical Oceanographic Real-Time System, (PORTS, US NOAA/NOS),	110
polar plot,	86
primary tide station,	64
principal axis currents,	55,87
Principal Components Analysis (PCA),	109
principle of superposition,	26
progressive, standing waves,	34
reduction in variance,	95
regression of lunar nodes,	21
residual current,	91
reversing currents,	86
RMS error,	95
rotary currents,	86
San Francisco Bay currents,	110
Saffir-Simpson hurricane scale,	135
sea level trend (secular trend),	65, 130
seasonal tides,	19, 56, 126
seiche (seiches),	33, 148
shallow water tides,	56, 59, 101, 102
shallow water wave (wavelength),	29, 76
significant wave height,	27
simple shallow-water wave model,	75
slack water (tidal),	87, 116
solar tides,	16
solar semidiurnal tide,	47
solar semiannual (annual) tide,	64, 126
spring-neap cycle,	17, 47, 48
static tide concepts,	13
station datum,	54
stilling well,	32
Stokes drift,	30
storm surge,	7, 66, 134
storm tide,	7, 66, 134
stratification, de-stratification,	6

summer (winter) solstice,	19
synodic month,	18
tidal benchmark,	63
tidal circulation,	5
tidal constituent (see harmonic constituent)	
tidal currents,	2, 55
tidal datum,	54
tidal epoch,	64
tidal form number,	51
tidal (non-tidal) flushing,	90
tidal prediction formula,	123, 125
tidal prism,	6, 90
tidal range,	2, 68
tide gauge,	32
tide staff,	62
tide mills,	3
tide producing force,	22
Tidewater region of Maryland, Virginia,	1
tractive forces,	9,12
tropic (equatorial) tides,	15
tsunami,	25, 134
United Kingdom Hydrographic Office, British Admiralty,	104, 123
universal time,	53
U,V currents,	109
vector averaging,	95
vernal (autumnal) equinox,	19
water level, time local,	142
wave attenuation (reflection),	76
wave celerity,	28
wave climatology,	27
wave dynamics (energy),	37
wave height (period, length),	26
wave shoaling (refraction),	28
wetlands	4